Machining Simulation

Using SOLIDWORKS CAM 2023

Kuang-Hua Chang, Ph.D.
School of Aerospace and Mechanical Engineering
The University of Oklahoma
Norman, OK

SDC
PUBLICATIONS

SDC Publications

P.O. Box 1334

Mission, KS 66222

913-262-2664

www.SDCpublications.com

Publisher: Stephen Schroff

ISBN-13: 978-1-63057-570-0

ISBN-10: 1-63057-570-4

Printed and bound in the United States of America.

Preface

In this book, *Machining Simulation using SOLIDWORKS CAM 2023*, we discuss the concept and steps in conducting machining simulation using SOLIDWORKS CAM (www.solidworks.com/product/solidworks-cam), which is a parametric and feature-based virtual machining software offered as an add-in to SOLIDWORKS. We present the important aspects of conducting machining simulations and help readers to create such simulations for support of practical machining tasks using SOLIDWORKS CAM. In order to provide a more comprehensive understanding of machining simulations, we discuss NC (numerical control) part programming and G-code verification, as well as introduce application case studies that involve bringing the G-code post processed by SOLIDWORKS CAM to HAAS (www.haascnc.com) CNC (computer numerical control) mill and lathe to physically cut parts. From the applications, we point out important lessons learned in transition from virtual to physical machining. In addition, we note that machining capabilities offered in the 2023 version of SOLIDWORKS CAM are somewhat limited mainly to support machining 2.5-axis features. Therefore, we introduce third-party CAM modules that are seamlessly integrated to SOLIDWORKS, including CAMWorks, HSMWorks, and Mastercam for SOLIDWORKS. All three offer advanced and powerful machining simulation capabilities, and are popular and widely accepted by industry and academia.

SOLIDWORKS CAM, powered by CAMWorks (www.camworks.com), supports users to integrate design and manufacturing in one application, connecting design and manufacturing teams through a common software tool that facilitates product design in 3D solid models. SOLIDWORKS CAM provides good capabilities for support of machining simulations in a virtual environment. Capabilities in SOLIDWORKS CAM allow users to select CNC machines and cutting tools, extract or create machinable features, define machining operations, and simulate and visualize machining toolpath. Moreover, the toolpath generated can be converted into NC part programs (also referred to as NC code, G-code, or M code) to machine functional parts as well as die or mold for part production. In most cases, the toolpath is generated in a so-called CL (cutter location) data format and then converted to G-code using respective post processors.

By carrying out machining simulation, the machining process can be defined and verified early in the product design stage. Some, if not all, of the less desirable design features in the context of part manufacturing, such as deep pockets, holes or fillets of different sizes, or cutting on multiple sides, can be detected and addressed while the product design is still being finalized. In addition, machining-related problems, such as undesirable surface finish, surface gouging, and tool or tool holder colliding with stock or fixtures, can be detected and eliminated before mounting a stock on a CNC machine at shop floor. In addition, manufacturing cost, which often constitutes a significant portion of the product cost, can be estimated using the machining time estimated in the machining simulation.

The book covers basic concepts and frequently used commands and options required for readers to advance from a novice to an intermediate level in using SOLIDWORKS CAM. Basic concepts and commands introduced include extracting machinable features (such as 2.5 axis features), selecting a machine and cutting tools, defining machining parameters (such as feedrate, spindle speed, depth of cut, and so on), generating and simulating toolpath, and post processing CL data to output G-code for support of physical machining. The concept and commands are introduced in a tutorial style presentation using simple but realistic examples. Both milling and turning operations are included.

One of the unique features of this book is the incorporation of the CL data verification by reviewing the G-code post-processed from the toolpath. This helps readers understand how the G-code is generated by using the respective post processors, which is an important step and an ultimate way to confirm that the toolpath and G-code generated are accurate and useful.

Example files are prepared for you to go over all lessons in this book. You may download them from the publisher's website: www.sdcpublications.com.

This book is intentionally kept simple. It primarily serves the purpose of helping readers become familiar with SOLIDWORKS CAM in conducting machining simulation for practical applications. This is not a reference manual of SOLIDWORKS CAM. You may not find everything you need in this book for learning SOLIDWORKS CAM. But this book provides you with basic concepts and steps in using the software, as well as discussions on the G-code generated. After going over this book, readers should gain a clear understanding in using SOLIDWORKS CAM for conducting machining simulations and should be able to apply the knowledge and skills acquired to carry out machining assignments and bring machining consideration into engineering design.

This book should serve well for self-learners. A self-learner should have a basic physics and mathematics background, preferably a bachelor or associate degree in science or engineering. We assume that readers are familiar with basic manufacturing processes, especially milling and turning. Details related to basic milling, turning, and hole making can be found in excellent textbooks, such as *Manufacturing, Engineering & Technology,* 6th ed., by Serope Kalpakjian and Steven R. Schmid. Familiarity with G-code is extremely important in learning virtual machining and transitioning from virtual to physical machining. Therefore, we discuss NC part programming and G-code verification in this book, although it is not the main focus of the book. In addition to the discussion provided in the book, we encourage readers to review NC programming books, for example, *Technology of Machine Tools,* 7th ed., by Krar, Gill, and Smid. If you are interested in understanding how the toolpaths are generated, i.e., the theory and behind-the-scenes computation, you may refer to books, such as *e-Design, Computer-Aided Engineering Design*, or *Product Manufacturing and Cost Estimating using CAE/CAE*, written by the author. And certainly, we expect that you are familiar with SOLIDWORKS part and assembly modes. A self-learner should be able to complete the fourteen lessons of this book in about fifty hours. An investment of fifty hours will advance readers from a novice to an intermediate user level, a wise investment.

This book also serves well for class instruction. Most likely, it will be used as a supplemental reference for courses like CNC Machining, Design and Manufacturing, Computer-Aided Manufacturing, or Computer-Integrated Manufacturing. This book should cover five to six weeks of class instruction, depending on the course arrangement and the technical background of the students. Some of the exercise problems provided at the end of individual lessons may take noticeable effort for students to complete. The author strongly encourages instructors or teaching assistants to go through those exercises before assigning them to students.

For those who desire to learn more about SOLIDWORKS CAM, you may find additional references on the computers where SOLIDWORKS CAM is installed. Several tutorial manuals can be found on your computer where SOLIDWORKS CAM is installed:

C:\Program Files\SOLIDWORKS Corp\SOLIDWORKS CAM\Lang\English\Manuals

with example files located at:

C:\Users\Public\Documents\SOLIDWORKS\SOLIDWORKS 2023\CAM Examples

Also, a few useful videos can be found on YouTube (reviewed on May 19, 2023):

www.youtube.com/watch?v=Dti5zbcJ5fl (Tech Tip - SOLIDWORKS CAM Getting Started)
www.youtube.com/watch?v=XQAVIV6v7-g (SOLIDWORKS CAM and CAMWORKS - What's New 2023 Webinar)
www.youtube.com/watch?v=nZFEXHgMmmo (SOLIDWORKS CAM – Top 10 Tips and Tricks)

and websites of technical consulting firms, such as:

Hawk Ridge Systems: hawkridgesys.com/solidworks/cam
GoEngineer: www.goengineer.com/products/solidworks/cam

Note that in order to go over all 14 lessons provided in this book, you must have SOLIDWORKS CAM Professional version installed on your computer. SOLIDWORKS CAM Standard version does not support assembly machining nor turning operations.

We hope you enjoy the experience of reading this book and learning SOLIDWORKS CAM to conduct machining simulations.

KHC
Norman, Oklahoma
May 28, 2023

Acknowledgements

My sincere appreciation is due to Mr. Stephen Schroff at SDC Publications for his encouragement and support for converting the book idea into reality. Without his support and the help of his staff at SDC Publications, this book would still be in its primitive stage.

Thanks are due to undergraduate students at the University of Oklahoma (OU) for their help in testing examples included in this book. They made numerous suggestions that improved clarity of presentation and found numerous errors that would have otherwise crept into the book. Their contributions are greatly appreciated.

Gratitude is also due to OU then-undergraduate students, Barry Bosnyak and Robert Kunkel, for their diligent effort in carrying out the project of machining the robotic forearm members that produced results included in Lesson 12. Appreciation is due to another then-OU project team, including Ryan Bodlak, Vishma Galge, and Wyatt Maney, for their help in completing the baseball machining project. Information extracted from their project report created the basic contents of Lesson 13. Their generosity in giving permission to include their work in this book is acknowledged and highly appreciated.

About the Author

Dr. Kuang-Hua Chang is a professor for the School of Aerospace and Mechanical Engineering at the University of Oklahoma (OU), Norman, OK. He received his diploma in Mechanical Engineering from the National Taipei Institute of Technology, Taiwan, in 1980; and M.S. and Ph.D. degrees in Mechanical Engineering from the University of Iowa in 1987 and 1990, respectively. Since then, he joined the Center for Computer-Aided Design (CCAD) at Iowa as a Research Scientist and shortly after was promoted as CAE Technical Area Manager. In 1997, he joined OU. He teaches mechanical design and manufacturing, in addition to conducting research in computer-aided modeling and simulation for design and manufacturing of mechanical systems.

His work has been published in 10 books and more than 150 articles in international journals and conference proceedings. He has also served as technical consultant to US industry and foreign companies, including LG-Electronics, Seagate Technology, etc. He has served as Associate Editor for two international journals: *Mechanics Based Design of Structures and Machines* and *Computer-Aided Design and Applications*.

Table of Contents

Lesson 1: Introduction to SOLIDWORKS CAM

1.1 Overview of the Lesson

SOLIDWORKS CAM (www.solidworks.com/product/solidworks-cam), powered by CAMWorks (www.camworks.com), is a parametric, feature-based virtual machining software offered as an add-in to SOLIDWORKS. Such an add-in module supports users to integrate design and manufacturing in one application, connecting design and manufacturing teams through a common software tool that facilitates product design in 3D solid model. By defining areas to be machined as machinable features, SOLIDWORKS CAM is able to apply more automation and intelligence into CNC (Computer Numerical Control) toolpath generation. This approach is more intuitive and follows the feature-based modeling concepts of computer-aided design (CAD) systems. Because of this integration, you can use the same user interface and solid models for design and later to create machining simulation. Such a seamless integration completely eliminates file transfers using less-desirable standard file formats such as IGES, STEP, SAT, or Parasolid. Hence, the toolpath generated is on the SOLIDWORKS solid model, not on an imported approximation. In addition, the toolpath generated is associative with SOLIDWORKS parametric solid model. This means that if the solid model is changed, the toolpath is changed automatically with minimal user intervention.

One unique feature of SOLIDWORKS CAM is the AFR (automatic feature recognition) technology. AFR automatically recognizes machinable features in solid models of native format or neutral file format; including mill features such as holes, slots, pockets and bosses; and turn features such as outside and inside diameter profiles, faces, grooves and cutoffs. This capability is complemented by interactive feature recognition (IFR) for recognizing complex multi-surface features, as well as creating contain and avoid areas.

Another powerful capability found in SOLIDWORKS CAM is its technology database, called TechDB™, which provides the ability to store machining strategies feature-by-feature, and then reuse these strategies to facilitate the toolpath generation. Furthermore, TechDB™ is a self-populating database, which contains information about the cutting tools and the machining parameters used by the operator. It also maintains information regarding the cutting tools available on the shop floor. This database within SOLIDWORKS CAM can be customized to meet the user's and the shop floor's requirements. This database helps in storing the best practices at a centralized location in support of machining operations, both in computers and at the shop floors.

We set off to discuss NC part programming and learn to use SOLIDWORKS CAM NC Editor and other NC viewers to review and verify the NC part program or G-code. With the basic understanding of NC part programming, we then start the main topic of the book: virtual machining simulation. We discuss the topic by exploring and learning capabilities offered by SOLIDWORKS CAM. The lessons and examples offered in this book are carefully designed and structured to support readers becoming effective in using

SOLIDWORKS CAM and becoming competent in carrying out virtual machining simulation for general machining applications.

We assume that you are proficient with part and assembly modeling capabilities in SOLIDWORKS, have some knowledge about NC part programming and the format and semantics of the G-code, and understand the practical aspects of setting up and conducting machining operations on CNC machines at the shop floor. Therefore, this book starts with a brief review on NC part programming so that readers will be able to read and revise, if necessary, the G-code generated from SOLIDWORKS CAM. We then focus on presenting virtual machining simulation using SOLIDWORKS CAM for toolpath and G-code generations. In addition to learning SOLIDWORKS CAM, we present two applications lessons, one for milling and one for turning operations, in which we load the G-code generated by SOLIDWORKS CAM to CNC mill and lathe, respectively, and carry out physical machining operations at the shop floors. In these two lessons, we discuss the important steps in transition from virtual machining to physical material cutting at the shop floor, point out numerous practical aspects of CNC machining, and verify the accuracy of the G-code post processed by SOLIDWORKS CAM.

Although SOLIDWORKS CAM is useful and powerful in support of most machining assignments, capabilities offered in the 2023 version are mainly for support of machining 2.5 axis features and are somewhat limited for machining freeform surfaces often found in the die and mold machining operations. There are other third-party CAM modules fully integrated into SOLIDWORKS that offer more capabilities than SOLIDWORKS CAM. At the end of the book (Lesson 14), we introduce three such modules, including CAMWorks (www.camworks.com), HSMWorks (www.autodesk.com/products/hsm/overview), and Mastercam for SOLIDWORKS (www.mastercam.com), to show readers alternatives available when it comes to machining parts that involve freeform surfaces and beyond.

1.2   Virtual Machining

Virtual machining is a simulation-based technology that supports engineers in defining, simulating, and visualizing machining operations in a computer environment using computer-aided manufacturing (CAM) tools, such as SOLIDWORKS CAM. Working in a virtual environment offers advantages of ease in making adjustments, detecting errors and correcting mistakes, and understanding machining operations through visualization of machining simulations. Once finalized, the toolpath can be converted to G-code and uploaded to a CNC machine at the shop floor to physically machine parts.

The overall process of using SOLIDWORKS CAM for creating machining simulation, as illustrated in Figure 1.1, consists of several steps: create design model (solid models in SOLIDWORKS part or assembly), choose NC machine and create stock, extract or identify machinable features, generate operation plan, generate toolpath, simulate toolpath, and convert toolpath to G-code through a post processor. Note that before extracting machinable features, we select an NC machine; i.e., a mill or lathe, choose a tool crib, select a suitable post processor, and create a stock.

The operation plan involves the NC operations to be performed on the stock, including selection of part setup origin, where G-code program zero is located. Also included is choosing cutting tools, defining machining parameters, such as feedrate, stepover, depth of cut, etc. Note that operation plans can be automatically generated by the technology database of SOLIDWORKS CAM when a machinable feature is extracted or created manually beforehand. Users may make changes to any part of the operation plan, for instance, choosing a different cutting tool (also called cutter or tool in the book), entering a different feedrate, adjusting depth of cut, and so on. After an operation plan is complete, SOLIDWORKS CAM generates toolpath automatically. Users may carry out material removal simulation, step through machining toolpath, and review important machining operation information, such as machining time that contributes largely to the manufacturing cost.

The design model (also called part, design part or target part in this book), which is a SOLIDWORKS part representing the perfectly finished product, is used as the basis and starting point for all machining simulations. Machinable features are extracted or created interactively on the design model as references for toolpath of individual operations. By referencing the geometry of the design model, an associative link between the design model and the stock is established. Because of this link, when the design model is changed, all associated machining operations are updated to reflect the change.

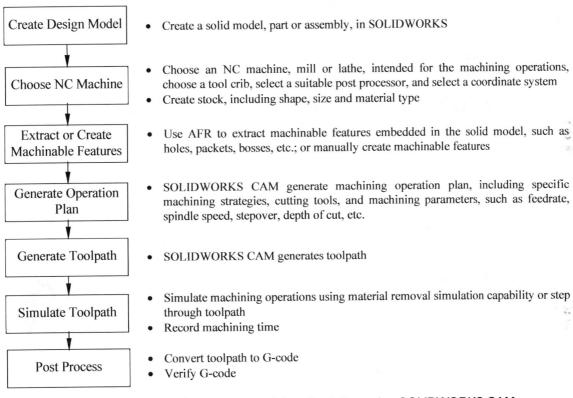

Figure 1.1 Process of creating machining simulation using SOLIDWORKS CAM

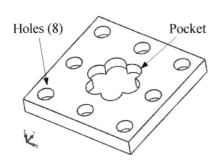

Figure 1.2 Design model in SOLIDWORKS

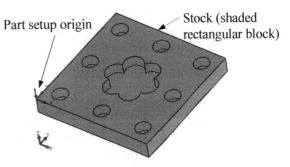

Figure 1.3 Stock enclosing the design model

The following example, a block with a pocket and eight holes shown in Figure 1.2, illustrates the concept of conducting virtual machining using SOLIDWORKS CAM. The design model consists of a base block (a boss extrude solid feature) with a pocket and eight holes that can be machined from a rectangular block (the raw stock shown in Figure 1.3) through pocket milling and hole drilling operations, respectively.

A generic NC machine *Mill-in.* (3-axis mill of inch system) available in SOLIDWORKS CAM is chosen to carry out the machining operations for this example. For example, the toolpath for machining the pocket (both rough and contour milling operations), as shown in Figure 1.4, can be generated referring to the part setup origin at the top left corner of the stock (see Figure 1.3). Users can step through the toolpath, for example, the contour milling operation for cutting the pocket with tool holder turned on for display, as shown in Figure 1.5. The material removal simulation of the same toolpath can also be carried out like that of Figure 1.6.

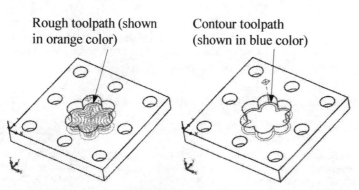

Rough toolpath (shown in orange color)   Contour toolpath (shown in blue color)

Figure 1.4 Toolpath of the pocket milling operations

Figure 1.5 Step through toolpath

1.3   SOLIDWORKS CAM Packages

SOLIDWORKS CAM offers four packages, Standard, Professional, Mechanist Standard, and Mechanist Professional. SOLIDWORKS CAM Standard supports you to quickly program individual milled parts and configurations without leaving the SOLIDWORKS 3D CAD environment. You have full access to defining rules within SOLIDWORKS CAM to create and build to your own standards. SOLIDWORKS CAM Standard also supports Tolerance-Based Machining (TBM) that automatically identifies and compensates for toleranced dimensions while generating machining toolpath.

SOLIDWORKS CAM Professional builds on the capabilities of SOLIDWORKS CAM Standard with increased programming capabilities. The additional features include High-Speed Machining (HSM), configurations, assembly machining, turning, and 3+2 programming to drive four- and five-axis machines[1]. Note that the Professional version also includes turning operations.

Figure 1.6 The material removal simulation

---

[1] Extracted from SOLIDWORKS website: www.solidworks.com/product/solidworks-cam

SOLIDWORKS CAM Mechanist Standard provides all the functionality found in SOLIDWORKS CAM Standard plus a part only modeling environment allowing users to work seamlessly with SOLIDWORKS part files and import several neutral file formats. The connectivity makes it easy to work with customers and vendors to collaborate on the manufacturing process.

SOLIDWORKS CAM Machinist Professional provides SOLIDWORKS parts and assemblies to allow the user to import and design fixtures and other manufacturing components used in the machining process. By using assembly mode in SOLIDWORKS CAM Professional, automatic toolpath clipping can be used to ensure the toolpath of the NC programs do not collide with custom fixtures or vises.

Please note that, as of 2023, SOLIDWORKS CAM offers 2½ axis (or 2.5 axis) milling capabilities. The 3+2 programming mentioned above is in fact quite misleading. 3+2 programming really should be called 2½ + 2. It is not 4 or 5 axis machining. All it does is use the 4th and 5th axis for positioning. They do not move once positioned, and all subsequent operations are really just 2½ axis motion. Users have to use other third-party CAM modules (please review Lesson 14), such as CAMWorks, HSMWorks, or Mastercam, for full three axis or more simultaneous axes of motion.

Overall, the machining modules included in SOLIDWORKS CAM Professional support 2.5 axis milling and 2-axis turning; that is:

- 2.5 axis mill: includes roughing, finishing, thread milling, face milling and single point cycles (drilling, boring, reaming, tapping) to machine prismatic features;

- 2 axis turning: includes roughing, finishing, grooving, threading, cutoff and single point cycles (drilling, boring, reaming and tapping).

All the above capabilities are discussed in this book and are illustrated using simple yet practical examples. In addition, SOLIDWORKS CAM supports machining of multiple parts in a single setup. Parts are assembled as SOLIDWORKS assembly, which includes parts, stock, clamps, fixtures, and jig table in a virtual environment that accurately represents a physical machine setup at shop floor. A multipart machining example, as shown in Figure 1.7, with ten identical parts in an assembly will be introduced in Lesson 7. Furthermore, machining features on multiple planes of parts mounted on the respective four faces of the tombstone, as shown in Figure 1.8, in a single setup is supported. More about multiplane machining operations can be found in Lesson 8.

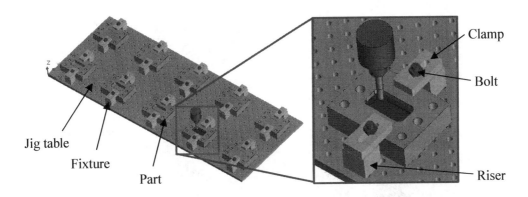

Figure 1.7 The material removal simulation of a multipart machining example

1.4   User Interface

The overall design of the SOLIDWORKS CAM user interface, as shown in Figure 1.9 that includes the layout and windows, buttons, menu selections, dialog boxes, etc., is identical to that of SOLIDWORKS CAD. SOLIDWORKS users should find it is straightforward to maneuver in SOLIDWORKS CAM. As shown in Figure 1.9, the user interface window of SOLIDWORKS CAM consists of pull-down menus, command buttons, graphics area, and feature manager window.

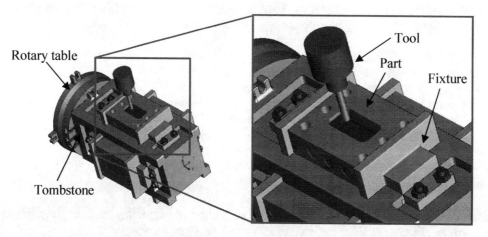

Figure 1.8 Material removal simulation of the multiplane machining example

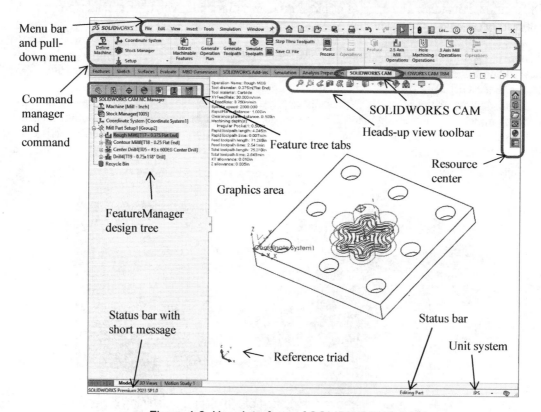

Figure 1.9 User interface of SOLIDWORKS CAM

Table 1.1 The major command buttons in SOLIDWORKS CAM

Button Symbol	Name	Function
Define Machine	Define Machine	Allows you to define the machine tool that the part will be machined on, such as 3-axis mill.
Coordinate System	Define Coordinate System	Allows you to define a coordinate system and assign it as the Fixture Coordinate System for the active machine.
Stock Manager	Stock Manger	Allows you to define the mill stock from a bounding box, an extruded sketch or an STL file.
Setup	Mill Part Setup	Allows you to create Mill Part Setups that define (1) tool orientation (or feed direction), (2) G-code program zero, and (3) the X direction of tool motion.
Extract Machinable Features	Extract Machinable Features	Initiates automatic feature recognition (AFR) to automatically extract solid features that correspond to the machinable features defined in the technology database (TechDB™). The types of machinable features recognized for mill and turn are different. SOLIDWORKS CAM determines the types of features to recognize based on the NC machine selected. The machinable features extracted are listed in the feature manager window under the SOLIDWORKS CAM feature tree tab .
Generate Operation Plan	Generate Operation Plan	Generates operation plan automatically for the selected machinable features. The operation plan and associated parameters are generated based on rules defined in TechDB™. An operation contains information on how the machinable features are to be machined. The operations generated are listed in the feature manager window under the SOLIDWORKS CAM operation tree tab .
Generate Toolpath	Generate Toolpath	Creates toolpath for the selected operation plan and displays the toolpath on the part. A toolpath consists of cutting entities (line, circle, arc, etc.) created by a machining operation that defines tool motion.
Simulate Toolpath	Simulate Toolpath	Provides a visual verification of the machining process for the current part by simulating the tool motion and the material removal process.
Step Thru Toolpath	Step Through Toolpath	Allows you to view toolpath movements either one movement at a time, a specified number of movements or all movements.
Save CL File	Save CL File	Allows you to save the current operation and associated parameters in the technology database as CL (cutter location) data for future use.
Post Process	Post Process	Translates toolpath and operation information into G-code for a specific machine tool controller.

An example file, *Lesson 1 with Toolpath.SLDPRT*, is prepared for you to browse numerous capabilities and become familiar with selections, buttons, commands and options of the SOLIDWORKS CAM user interface. This file (and all example files of the book) is available for download at the publisher's website (www.sdcpublications.com). You may review Section 1.5 for the detailed steps to bring the example into SOLIDWORKS CAM.

The graphics area displays the solid model and machining toolpath on the solid model with which you are working. The pull-down menus provide basic solid modeling functions in SOLIDWORKS and machining functions in SOLIDWORKS CAM. The command buttons of SOLIDWORKS CAM tab above the graphics area offer all the functions required to create and modify virtual machining operations in a generic order. Major buttons include extract machinable features, generate operation plan, generate toolpath, simulate toolpath, step through toolpath, and save CL file. When you move the mouse pointer over these buttons, a short message describing the function will appear. Some of the frequently used buttons in SOLIDWORKS CAM and their functions are summarized in Table 1.1 for your reference.

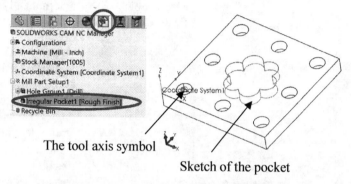

The tool axis symbol

Sketch of the pocket

Figure 1.11 Selecting a machinable feature under the SOLIDWORKS CAM feature tree tab

Figure 1.10 Solid features listed in the feature manager window under the FeatureManager design tree tab

Figure 1.12 SOLIDWORKS CAM operation tree tab

Figure 1.13 SOLIDWORKS CAM tools tree tab

There are four feature tree tabs on top of the feature manager window that are highly relevant in learning SOLIDWORKS CAM. The leftmost tab, FeatureManager design tree 🍁 (see Figure 1.10), sets the display to the SOLIDWORKS design tree (also called model tree, solid feature tree, or SOLIDWORKS browser), which lists solid features and parts created in SOLIDWORKS part and assembly in the feature manager window.

The third tab from the right, SOLIDWORKS CAM feature tree tab 📑 (see Figure 1.11), shifts the display to the SOLIDWORKS CAM feature tree, which lists machinable features extracted or interactively created from the solid model. The tree initially shows only the *Configurations*, *Machine* (for example, *Mill-inch* in Figure 1.11), *Stock Manager*, *Coordinate System*, and *Recycle Bin*. The *Machine* node indicates the current machine chosen. You will have to select the correct machine before you begin working on a part. If you click any machinable feature, an outline view of the machinable feature appears in the part in the graphics area. For example, the sketch of the pocket appears when clicking *Irregular Pocket1* in the feature tree, as shown in Figure 1.11. Note that a symbol 🔧 (called tool axis symbol) appears indicating the tool axis direction (or feed direction) of the current mill part setup.

The second tab from the right, SOLIDWORKS CAM operation tree 🖳 (shown in Figure 1.12), shifts the display to the SOLIDWORKS CAM operation tree. After you select the *Generate Operation Plan* command, the operation tree lists machining operations of the respective machinable features.

Similar to SOLIDWORKS, right clicking a machining operation in the operation manager tree will bring up command options that you can choose to modify or adjust the machining operation, such as feedrate, spindle speed, and so on. Clicking any operations after selecting the *Generate Toolpath* command will bring up the corresponding toolpath displayed on the part in the graphics area, like that of Figure 1.4.

The rightmost tab, SOLIDWORKS CAM tools tree 🔧 (shown in Figure 1.13), changes the display to SOLIDWORKS CAM tools tree. SOLIDWORKS CAM tools tree lists tools available in the tool cribs chosen for the machining task.

1.5 Opening Lesson 1 Model and Entering SOLIDWORKS CAM

A virtual machining model for the simple block example shown in Figure 1.2 has been created for you. You may download all example files from the publisher's website, unzip them, and locate the model files under Lesson 1 folder. Copy or move Lesson 1 folder to your hard drive.

Start SOLIDWORKS, and open part file: *Lesson 1 with toolpath.SLDPRT*. You should see a solid model like that of Figure 1.2 appears in the graphics area.

Entering SOLIDWORKS CAM from SOLIDWORKS is straightforward. You may simply click the SOLIDWORKS CAM feature tree tab 📑 or operation tree tab 🖳 to browse respective machining nodes. You may right click any node listed in the feature tree or operation tree to review or modify the machining model. You may also choose options under the pull-down menu *Tools > SOLIDWORKS CAM* to launch the same commands as those listed in Table 1.1 (and more) that support you to extract machinable features, generate an operation plan, and so on.

If you do not see the SOLIDWORKS CAM feature tree or operation tree tab, you may have not activated the SOLIDWORKS CAM add-in module. To activate the SOLIDWORKS CAM module, choose from the pull-down menu

Tools > Add-Ins

In the *Add-Ins* dialog box shown in Figure 1.14, click *SOLIDWORKS CAM 2023* in both boxes (*Active Add-ins* and *Start Up*) and then click *OK*. You should see that SOLIDWORKS CAM tab appears above the graphics area like that of Figure 1.9 and the three SOLIDWORKS CAM tree tabs (feature, operation, and tool) added to the top of the feature manager window.

If you still do not see SOLIDWORKS CAM tab above the graphics area, you may need to restart SOLIDWORKS to activate newly selected SOLIDWORKS CAM modules. Before going over this tutorial lesson, you are encouraged to check with your system administrator to make sure SOLIDWORKS and SOLIDWORKS CAM have been properly installed on your computer. Note again that you will need to have SOLIDWORKS CAM Professional version to go over all fourteen lessons provided in this book.

Another point worth noting is that the auto saving option might have been turned on in SOLIDWORKS CAM by default. Often, it is annoying to get interrupted by this auto saving function every couple minutes asking if you want to save the model. You may turn this auto save option off by choosing from the pull-down menu

Tools > SOLIDWORKS CAM > Options

In the *Options* dialog box (Figure 1.15), select *Disable Auto Saving* under the *General* tab to turn it off. Then click *OK*.

To browse an existing SOLIDWORKS CAM model, you may click any machinable features listed under the SOLIDWORKS CAM feature tree tab 🔧 to display the feature in the graphics area. For example, the pocket profile sketch like that of Figure 1.11 appears in the solid model after clicking *Irregular Pocket1*.

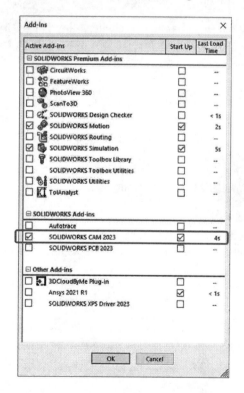

Figure 1.14 The *Add-Ins* dialog box

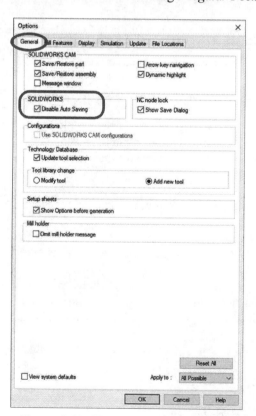

Figure 1.15 The *Options* dialog box

You may also click any machining operation under SOLIDWORKS CAM operation tree tab ▣ to display the toolpath of the selected operation. For example, clicking *Rough Mill* or *Contour Mill* displays toolpath on the solid model like those of Figure 1.4. You may step through the toolpath of an operation (for example, see Figure 1.5) by right clicking it and choosing *Step Through Toolpath*. You may right click an operation and choose *Simulate Toolpath* to simulate a material removal process of the operation like that of Figure 1.6.

1.6   Extracting Machinable Features

Machining operation plans (or simply operations) and toolpaths can be generated only on machinable features. A unique and appealing technical feature in SOLIDWORKS CAM is the automatic feature recognition (AFR) capability, which analyzes the solid features in the part solid model and automatically extracts mill features such as holes, slots, pockets and bosses; and turn features such as outside and inside diameter profiles, faces, grooves and cutoff. The AFR capability helps in reducing the time spent by the designer to feed in data related to creating machining simulation.

A set of machinable features for milling and turning operations that can be extracted by AFR are summarized in Appendix A. The associated machining strategies of individual machinable features can be found in Appendix B.

The *Extract Machinable Features* command ⬚ Extract Machinable Features initiates AFR. Depending on the complexity of the part, AFR can save considerable time in extracting 2.5 axis features, such as holes, pockets, slots, bosses, etc., either prismatic (with vertical walls) or tapered.

AFR cannot recognize every single feature on complex parts and does not recognize features beyond 2.5 axis. To machine these areas, you need to define machinable features manually or interactively using the interactive feature recognition (IFR) wizard. For example, you may define a *Multi Surface* feature manually by selecting faces to be cut and faces to avoid in the design model. More on this topic will be discussed in this book, for example, in Lesson 2: Simple Plate, and Lesson 6: Freeform Surface.

1.7   Technology Database

SOLIDWORKS CAM technology database, TechDB™, is a self-populating database which contains all the information about the machine, cutting tools and the parameters used by the operator, and rules of repetitive NC operations (called machining strategy or strategy in SOLIDWORKS CAM) for the respective machinable features. This database within SOLIDWORKS CAM can be customized to meet the user's and the shop floor's requirements. This database helps keep the best practices at a centralized location in the tool room; thus, it eliminates the non-uniformity in practicing virtual and physical machining operations.

Using a set of knowledge-based rules, SOLIDWORKS CAM analyzes the machinable features and selects machining strategies to machine machinable features extracted or defined in the design model. In this approach, features are classified according to the number of possible tool approaching directions that can be used to machine them. The knowledge-based rules are applied to assure that users are provided with desired cutting operations. The rules that determine machining operations for a respective machinable feature can be found in Appendix B, for both milling and turning operations.

The technology database is shipped with data that is considered generally applicable to most machining environments. Data and information stored in the database can be added, modified, or deleted to meet the user's specific needs in practice.

1.8   Tutorial Examples

There are fourteen lessons included in this book. In addition to the example of the current lesson, we go over two lessons discussing NC part programming before resuming our discussion on more machining simulation lessons. In Lesson 2, we discuss (or more like a review) NC part programming, in which we review NC address codes, preparatory functions, and auxiliary functions, before going over examples of pocket milling, trajectory milling, profile milling, and canned cycle operations. In Lesson 3, we introduce SOLIDWORKS CAM NC Editor, in which we learn to use the Editor to review and verify G-code. We also briefly go over a few NC reviewers available on-line.

The next eight lessons, Lessons 4 to 11, illustrate step-by-step details of creating machining operations and simulating toolpath capabilities in SOLIDWORKS CAM. We start in Lesson 4 with a simple plate example, which provides you with a quick run-through for creating a contour profile mill operation using a 3-axis mill.

Lessons 5 through 9 focus on milling operations. We include examples of machining 2.5 axis features using a 3-axis mill in Lesson 5, machining a freeform surface of a solid feature in Lesson 6, machining a set of identical parts in an assembly in Lesson 7, and machining features of parts mounted on multiple planes using a 3-axis mill with a rotary table in Lesson 8. In Lesson 9, we discuss tolerance-based machining (TBM), which leverages SOLIDWORKS dimensions, tolerance ranges and surface finish annotations to select machining strategies for operations and machine parts to the mean of asymmetric tolerances.

Lessons 10 and 11 focus on turning operations. In Lesson 10, we use a simple stepped bar example to learn basic capabilities in creating turning operations and understanding G-code generated by SOLIDWORKS CAM. In Lesson 11, we machine a similar example with more turn features to gain a broader understanding of the turning capabilities offered by SOLIDWORKS CAM.

Lessons 12 and 13 discuss transition from virtual to physical machining, which is no small matter. In Lesson 12, we present an application extracted from a student project that involves milling operations for machining a robotic rover forearm member. SOLIDWORKS CAM was employed to conduct virtual machining and toolpath generation for machining not only the forearm member but also a custom fixture. G-code generated by SOLIDWORKS CAM is then uploaded to a HAAS CNC mill to machine the fixture and the part. In Lesson 13, we discuss a turning application, in which we turn a scaled baseball bat on a HAAS CNC lathe using G-code created from SOLIDWORKS CAM. The goal of these two lessons is to offer readers a flavor of the role that SOLIDWORKS CAM is able to play in practical machining assignments. Moreover, we point out important factors for you to consider before transporting the results of virtual machining to CNC machines at the shop floor to physically cut the part.

In Lesson 14, we introduce third-party CAM modules that are fully integrated with SOLIDWORKS, including CAMWorks, HSMWorks, and Mastercam for SOLIDWORKS. In particular, we discuss milling operations involving full three axis or more simultaneous axes of motion, which are not supported in SOLIDWORKS CAM 2023 version. Note that you may be able to download trial versions of HSMWorks and Mastercam for SOLIDWORKS, install them on your computer, and go over Lesson 14.

One thing we emphasize in this book is the verification of the G-code converted from machining simulation. Learning the menu selections and button clicking of SOLIDWORKS CAM for generating machining simulation is important. On the other hand, the simulation must lead to something useful at the shop floor. That is, the G-code converted from machining simulation must be accurate and compatible with the NC machines at the shop floor. The G-code must be ready and able to produce parts as desired without major hurdles. Please note that no software is 100% error-proof and bug-free. Therefore, it is extremely important that we carefully review and verify the G-code before loading it to the CNC machine for material cutting. At the end of most lessons, we review and verify the G-code converted. Readers are strongly encouraged to do the same while applying the skills learned from this book to their own machining projects.

All examples and topics to be discussed in individual lessons are summarized in Table 1.2.

Table 1.2 Examples employed and topics to be discussed in this book

Lesson	Example	Machining Model	Problem Type	Topics
2	Trajectory milling, profile milling, pocket milling, hole drilling		Manual NC part programming	1. NC programming, 2. G-code and M code, 3. Tool radius compensations, 4. Canned cycles, subroutines, and more.
3	Trajectory milling, profile milling, hole drilling, turning operation		Review and verify G-code using SOLIDWORKS NC Editor	1. User interface of SOLIDWORKS NC Editor, 2. Backplotter, tool setup, coordinate system, and more, 3. Tool radius compensation, 4. Turning operations, 5. Toolpath simulation, 6. Other G-code viewers and editors.
4	Simple Plate		3-axis contour (or profile) milling	1. A first experience of using the capabilities of SOLIDWORKS CAM 2023, 2. A complete process of using SOLIDWORKS CAM to create a milling operation from the beginning all the way to the post process that generates G-code, 3. Extracting machinable feature using interactive feature recognition (IFR), 4. Reviewing G-code generated.
5	2.5 Axis Features		3-axis pocket milling and hole drilling	1. Extracting 2.5 axis machinable features using automatic feature recognition (AFR), 2. Manually creating machinable feature for face milling operation, 3. Defining part setup origin for G-code generation, 4. Modifying machining parameters to regenerate toolpath, 5. Reviewing G-code generated.

Table 1.2 Examples employed and topics to be discussed in this book (cont'd)

Lesson	Example	Machining Model	Problem Type	Topics
6	Freeform Surface		Machine a freeform surface using 3-axis mill	1. Creating a multi-surface machinable feature and select avoid surface feature to restrain area of toolpath generation, 2. Creating *Area Clearance* (rough cut) and *Z Level* (finish cut) operations, 3. Creating section views for a closer look at the freeform surface in material removal simulation, 4. Limitations of SOLIDWORKS CAM for machining sculpture surface.
7	Multipart Machining		Machine a set of identical parts in an assembly	1. Select fixture coordinate system for cutting multiple parts, 2. Creating instances of part for machining operations, 3. Defining stocks for individual instances, 4. Selecting components in the assembly for the tools to avoid, 5. Simulating material removal for cutting multiple parts, 6. Reviewing G-code generated.
8	Multiplane Machining		Machine features on multiple planes using a 3-axis mill with a rotary table	1. Machining parts with machinable features on multiple planes, 2. Setting a rotation axis of the rotary table as the 4th axis, 3. Selecting components to be included in material removal simulation, 4. Rotating tool vs. rotating stock in material removal simulation, 5. Reviewing G-code generated.
9	Tolerance-Based Machining		Operations and toolpaths generation based on tolerances	1. Defining dimension tolerances using *DimXpert*, 2. Tolerance-based machining settings, 3. Running a TBM simulation, 4. Result interpretation.
10	Turning a Stepped Bar		Basic turning operations using 2-axis lathe	1. A complete process in using SOLIDWORKS CAM to create a turning simulation from the beginning all the way to the post process that generates G-code, 2. Manually creating machinable features, 3. Reviewing G-code generated.

Table 1.2 Examples employed and topics to be discussed in this book (cont'd)

Lesson	Example	Machining Model	Problem Type	Topics
11	Turning a Stub Shaft		Advanced turning operations	1. Extracting machinable features for turning, including face, groove, thread, and holes at both ends using AFR, 2. Manually creating turn thread feature, generating toolpath, 3. Reviewing G-code generated.
12	Machining Robotic Forearm Members		Practical application for milling operations	1. Solid model of the forearm member, 2. Machining strategy, 3. Custom fixture and stock-in-progress, 4. HAAS mill setup and stock installation, 5. G-code generated by SOLIDWORKS CAM, 6. Finished part.
13	Turning a Baseball Bat		Practical application for turning operations	1. Solid model of a scaled baseball bat, 2. Turning operations, 3. HAAS lathe setups and round stock installation, 4. G-code generated by SOLIDWORKS CAM, 5. Finished part.
14	Freeform surface example of Lesson 6		Freeform surface milling of 3-axis and multiaxis, and Machine Simulation	1. Third-party CAM modules for Multiaxis Surface Machining, 2. CAMWorks 5-axis surface milling operation and Machine Simulation in a setup with a tilt rotary table, 3. Volume milling, local milling, and surface milling using HSMWorks, 4. Volume milling, local milling, and multi-axis surface milling using Mastercam for SOLIDWORKS, plus Mastercam Simulator of a setup with a tilt rotary table.

[Notes]

Lesson 2: NC Part Programming

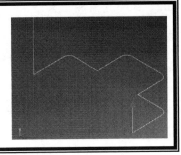

2.1 Overview of the Lesson

In this lesson, we go over a short review on NC part programming (or NC programming) for readers who may not be familiar with the topic. The goal is not to make you proficient in NC programming; instead, the goal is to ensure you acquire enough knowledge of NC programming so that you are able to read the G-code (also called M-code, G&M-code, NC code, NC program, or NC part program) generated by SOLIDWORKS CAM and modify the code if needed to meet specific requirements of CNC machines at the shop floor. Please note that, depending on the post processors installed on your computer, the G-code generated from SOLIDWORKS CAM (or other CAM software) may not be 100% compatible with your CNC machines. Some modifications are often required. Those who are interested in learning more about NC programming may refer to excellent references, such as *Technology of Machine Tools,* 7th ed., by Krar, Gill, and Smid.

NC part programming creates NC program. NC program is a language in which operators tell CNC (Computer Numerical Control) controller how to carry out a machining task. The NC program is often referred to as G-code, M-code, or G&M-code because the majority of the words (called NC words) of this language begin with the letters G or M. G words instruct a machine controller to tell the cutter where to move, how fast to move, and what path to follow. In addition, M words (for miscellaneous functions or machine functions) are auxiliary commands that instruct the machines to perform miscellaneous functions, such as spindle on/off, coolant on/off, feedrate, tool change, and so on.

Most CNC machines have a vocabulary of at least a hundred words, but only about thirty are used often. These thirty or so words are best memorized because they appear in almost every NC program and knowing them helps you work more efficiently.

An NC program can be composed manually, referred to as manual programming. NC program can also be generated by creating machining simulation using computer-aided manufacturing (CAM) software, such as SOLIDWORKS CAM. This is referred to as the CAD/CAM approach. The toolpath generated by CAM software can be converted to NC programs through post-processing. We will learn post-processing G-code from SOLIDWORKS CAM, starting in Lesson 4.

In this lesson, we briefly go over manual programming. We then provide a more in-depth discussion on the CAD/CAM approach for the remainder of the book. Before reviewing NC part programming, we will provide a brief discussion on NC machines and coordinate systems. We assume that readers are familiar with basic manufacturing processes, especially milling and turning. Details related to basic machining, including milling, turning, and hole making, can be found in excellent textbooks, such as *Manufacturing, Engineering & Technology,* 6th ed., by Serope Kalpakjian and Steven R. Schmid.

2.2   Basics of NC Machines

The major difference between NC machines and conventional machines is the way in which the various functions and cutter movements are controlled. In conventional machines, these are controlled by shop mechanists. In NC, these motions are controlled by the machine control unit (MCU), as depicted in Figure 2.1.

The MCU (brain of the NC) consists of a DPU (data processing unit) and a CLU (control loop unit). DPU reads the NC part program and decodes the program statements, processes the decoded information, and passes information to the CLU. The information includes position of each axis of the machine, its direction of motion, feedrate, and auxiliary function control signals (e.g., coolant on or off). CLU receives data from the DPU, converts them to control signals, and controls the machine via actuation devices that replace the hand wheel of the conventional machine. An actuation device could be a servomotor, a hydraulic actuator, or a step motor. A servomotor (or servo) is an electromechanical device in which an electrical input determines the position of the armature of a motor. Servos are used extensively in robotics and radio-controlled cars, airplanes, and boats.

The MCU gives instructions to the servo system, monitors both the position and velocity output of the system, and uses this feedback to compensate for errors between the program command and the system response. The instructions given to the servos are modified according to the measured response of the system. This is referred to as closed-loop control.

Each axis of motion is equipped with a driving (actuation) device. The primary three axes of motion are referred to as the X-, Y-, and Z-axes. They form the machine tool coordinate system. The XYZ system is a right-hand system and the location of its origin may be fixed (in older machines) or adjustable (floating zero) on modern machines.

The Z-axis is the most important axis for machining. This axis is always aligned with the spindle that imparts power, as shown in Figure 2.2(a) and (b). The spindle may rotate a stock such as in a lathe or it may rotate a tool as in a mill. Usually, the direction that moves away from the stock is defined as positive.

On a stock-rotating machine (e.g., lathe), the X-axis is the direction of tool movement, and a motion along its positive direction moves the tool away from the stock. On a milling or drilling machine, the positive X-axis points to the right when the programmer is facing the machine; see Figure 2.2(a). Y-axis is determined by X- and Z-axes through the right-hand rule. W-axis (less common) is parallel to Z-axis, but points in the opposite direction.

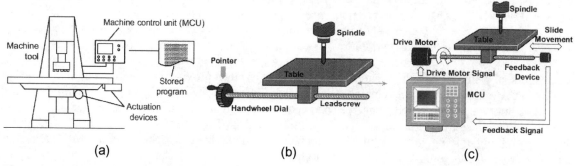

(a)   (b)   (c)

Figure 2.1   Configuration of an NC machine: (a) the machine control unit, (b) hand wheel dial, and (c) closed-loop control.

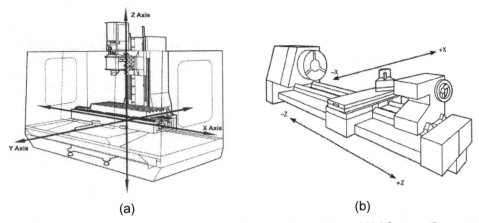

(a)   (b)

Figure 2.2 XYZ coordinate system, (a) on a mill (figure extracted from HAAS operation manual), and (b) on a lathe

2.3   NC Words and NC Blocks

NC programs list instructions to be performed in the order they are written. They read like a book, left to right and top down. Each sentence in an NC program is written on a separate line, called a *block*. Blocks are arranged in a specific sequence that promotes safety, predictability and readability, so it is important to adhere to a standard program structure.

All actions must be specified in the NC program as a set of blocks (typically, hundreds or thousands of blocks in an NC program). A block is a group of NC words terminated by the end-of-block character (usually a carriage return, CR). An NC word is composed of an alphabet address code (also referred to as letter address, address code, or identifier) followed by a numeric, integer or floating point. Figure 2.3 shows a sample block, which performs a cut from current cutter position to X = 2.35in. and Y = –0.475in. along a straight line at a feedrate of 5.0 in./min. (assuming inch is chosen for length unit).

In this sample block, there are seven NC words. N100 is simply a sequence number of the block. G01 and G41 are NC words that prepare cutter motion, where G01 specifies a linear motion and G41 turns on cutter radius compensation from left. X2.35 and Y–0.475 specify the next cutter location. F5.0 specifies the feedrate, and M07 is a miscellaneous function, which turns flood coolant on.

The X and Y commands specify the cutter location (CL), which locate the center point of the bottom face of an end mill or tip of a drill bit or a ball mill. Collectively, CL data form the toolpath, either connected by a straight line (G01) or circular arcs (G02/03).

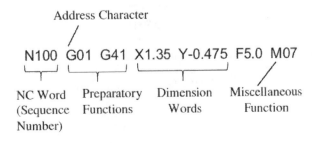

Figure 2.3 A sample NC block

2.4   Alphabet Address Codes

As seen above, an NC word starts with an alphabet address code. Shown in Table 2.1 are the address codes seen most frequently throughout an NC program. Some address codes are used only in milling or only in turning; most are used in both. Most address codes are common for most CNC machines. Some are assigned specific functions for a particular controller. Among the address codes, G and M are seen most frequently throughout NC programs. Therefore, NC program is also referred to as G-code, M-code, or G&M-code.

The G address code is used to specify the type of tool motion to occur, which is commonly referred to as preparatory functions, in the block. For example, as seen in the sample NC block of Figure 2.3, G01 specifies a linear motion and G41 turns on cutter radius compensation from left. An NC word with a prefix G is commonly referred to as G word. As mentioned before, NC part programs are often referred to as G-code since most NC words in an NC part program are G words. In addition to G01 and G41, there are more than 100 G words that specify the respective cutter motions. More discussion on G words is given in Section 2.5.

M address code is used to specify the miscellaneous machine functions, such as M07 that turns coolant on, as seen in the sample NC block of Figure 2.3. An NC word with a prefix M is commonly referred to as M word. Only one M word is allowed per NC block. M word is performed at the end of the block.

To offer the reader a more complete picture, a complete list of alphabet address codes with more detailed explanations for HAAS machines is given in Appendix C.

Table 2.1 A list of frequently used address codes

Code:	Function:
D	Tool diameter selection
F	Assigns a feedrate
G	Preparatory functions
H	Height offset register number
I	X axis incremental location of arc center
J	Y axis incremental location of arc center
K	Z axis incremental location of arc center
L	Loop count
M	Miscellaneous or machine functions
N	Block number (specifies the start of a block)
O	Program name
P	Dwell time, scaling
R	Retract distance used with G81, 82, 83 Radius when used with G02 or G03
S	Sets the spindle speed
T	Specifies the tool to be used
X	X axis coordinate
Y	Y axis coordinate
Z	Z axis coordinate

2.5   Preparatory Functions or G Words

Note that NC words with identifier G, or G words, prepare the MCU for a given operation, typically involving a cutter motion. A list of G words commonly found on FANUC[1] and similarly designed controls for milling and turning can be found in Table 2.2.

G words are assigned into groups. For example, G01 is a Group 1 command and G12 is in Group 0 for HAAS machines. Some G words are modal, some are non-modal. For example, G words in Group 0 of HAAS machines are non-modal; that is, they specify a function applicable to this block only and do not affect other blocks. All the other groups are modal, and the specification of one G word in the group cancels the previous G word applicable from that group. A modal G word applies to all subsequent blocks, so those blocks do not need to re-specify the same G word. More than one G word (from respective groups) can be placed in a block in order to specify all of the setup conditions for an operation.

A complete list of G words and their respective groups for HAAS CNC machines can be found in Appendix D. In this section, we discuss G00, G01, G02/03, G12/13, and canned cycle such as G81. Tool compensation commands G41-G43 will be discussed in Section 2.8.

Table 2.2 FANUC style G words

Code:	Function:
G00	Rapid motion, point to point, positioning, non-cutting
G01	Linear interpolation motion
G02	Circular interpolation motion, CW
G03	Circular interpolation motion, CCW
G04	Dwell
G12	Circular pocket milling, CW
G13	Circular pocket milling, CCW
G17	XY plane selection
G18	ZX plane selection
G19	YZ plane selection
G20	Select inches
G21	Select metric
G28	Return to home position (machine zero, aka machine reference point)
G30	Return to secondary home position
G40	Cutter radius compensation, cancel
G41	Cutter radius compensation, from left
G42	Cutter radius compensation, from right
G43	Tool length compensation
G54-59	Select work coordinate system 1-6
G80	Canned cycle cancel
G81-89	Drill canned cycles
G90	Absolute dimension programming
G91	Incremental dimension programming
G98	Initial point return
G99	R point return

---

[1] FANUC is a group of companies, principally FANUC Corporation of Japan; Fanuc America Corporation of Rochester Hills, Michigan, USA; and FANUC Europe Corporation S.A. of Luxembourg, that provide automation products and services such as robotics and CNC systems. FANUC is one of the largest makers of industrial robots in the world. FANUC had its beginnings as part of Fujitsu developing early numerical control (NC) and servo systems. The company name is an acronym for Fuji Automatic NUmerical Control.

G00: Rapid Motion

G00 moves the cutter from current position to the next position defined by the X, Y, and Z words. G00 is non-cutting, and provides rapid cutter motion and cutter retract.

Format: N_ G00 X_ Y_ Z_

Example:

N01 G00 X20. Y30. Z1.

This NC block instructs the controller to move the cutter from its current position to (20.0, 30.0, 1.0) if absolute programming G90 is in place; or by an increment of (20, 30, 1) if incremental programming G91 is turned on. The exact path on how to move from one point to another is determined by the machine controller. If the cutter current position is (5, 5, 1) or Point A shown in Figure 2.4(a), G00 X20. Y30. Z1. moves the cutter to Point C (20, 30, 1) following the dotted straight path AC or through a third point B determined by the controller.

G01: Linear Interpolation Motion

G01 moves cutter from current position to the next position defined by the X, Y, and Z words following a straight path. G01 word is a cutting command with a cutting feedrate specified by the F command.

Format: N_ G01 X_ Y_ Z_ F_

Example:

N100 G01 X30.0 Y20.0 F3.0

This NC block instructs the controller to perform a cutting motion by moving the cutter from the current position to (30.0, 20.0) with a feedrate 3.0. If the current position is Point A (5, 5) the toolpath of cutter motion is a straight line from Point A to Point C, as shown in Figure 2.4(b).

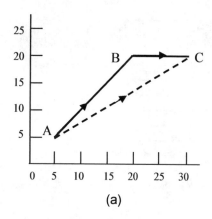
(a)

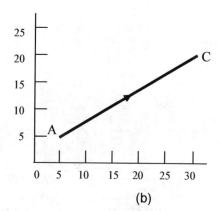
(b)

Figure 2.4 Toolpath of cutter motion, (a) G00 rapid motion, and (b) G01 linear interpolation cutting motion

G02/03: Circular Interpolation Motion, Clockwise/Counterclockwise

G02/03 instructs the controller to move the cutter along a circular path, a cutting motion followed by a feedrate defined by an F command. G02 moves the cutter in a clockwise direction, and G03 is counterclockwise, viewed from the positive Z-direction.

Format: N_ G02 X_ Y_ Z_ I_ J_ K_ F_

Example:

N10 G02 X20.0 Y10.0 I–5.0 J–15.0 F2.5

X, Y, and Z coordinates define the end point of the arc. I, J, and K values represent the incremental distance from the current position to the center of the arc.

If the current cutter position is at Point A (10, 20), as shown in Figure 2.5, the NC block above moves the cutter from Point A to Point B (20, 10) following a clockwise circular path with center point C (5, 5) and radius 15.81; i.e., $\sqrt{5^2 + 15^2}$, with a feedrate 2.5.

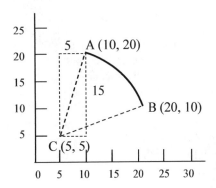

Figure 2.5 Toolpath of circular interpolation

Note that instead of using I and J words to determine the center point of the circular arc, R word can be used to specify the radius of the circular path for G02/03 commands.

For example, the two NC blocks, G00 X3.0 Y4.0 and G02 X–3.0 R5.0, moves the cutter to Point A (3, 4), and cuts a shorter circular arc of radius 5 to Point B (–3, 4), as shown in Figure 2.6(a). G02 X–3.0 R–5.0 moves the cutter from Point A to Point B following the path of a longer circular arc of radius 5, as shown in Figure 2.6(b).

G12/13: Circular Pocket Milling, Clockwise/Counterclockwise

G12 and G13 provide for pocket milling of a circular shape. They are different only in which direction of rotation is used. G12 is clockwise, and G13 is counterclockwise.

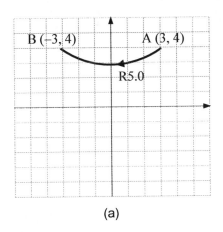

(a)

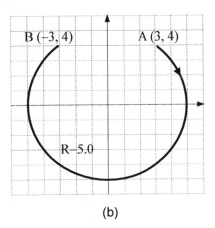

(b)

Figure 2.6 Toolpath of circular interpolation, (a) G02 with a positive R leading to a shorter arc, and (b) G02 with a negative R leading to a longer arc

Format: N_ G12 I_ K_ L_ Q_ F_ Z_

I: Radius of the first circle (or finished circle if no K specified)
K: Radius of the finished circle
L: Loop count for repeating deeper cuts
Q: Radius increment (must be used with K)
F: Feedrate
Z: Depth of cut

Note that G12/G13 is usually programmed together with tool radius compensation commands G41/42 so that a pocket of prescribed diameter can be machined using different size cutters. To turn on tool radius compensation, address code D is used to select a pre-registered cutter diameter. More about tool radius compensation will be discussed in Section 2.8. In this section, we assume no tool radius compensation.

The tool must be positioned at the center of the circle either in a previous block or in the current block using X and Y. The cut is performed entirely with circular motion of varying radii. To remove all the material within the pocket, one will have to specify an I and Q value less than tool diameter and a K value equal to the radius of the pocket. The radius of the actual pocket is K plus tool radius since we assume no tool radius compensation is turned on.

If no K and Q values are specified, K is assumed to equal I.

G12 (group 0) is non-modal.

If G91 (incremental) is specified and a positive L count is included, the Z increment is repeated L times at the F feedrate.

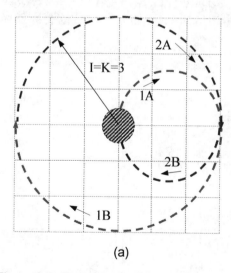
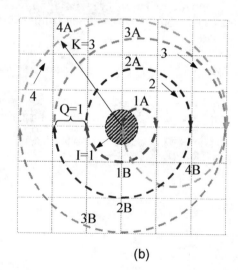

| (a) | (b) |

Figure 2.7 Toolpath of circular pocket milling, (a) G12, I = K = 3, and (b) G12 with I = 1, Q = 1, and K = 3

Example A: G12 I3. F10. Z–.25

If the current position of the cutter is (3, 3), as shown in Figure 2.7(a), this NC block moves the cutter along a circular path 1A of radius 1.5 clockwise, followed by a larger circle 1B and 2A of radius I = K = 3 (a complete 360° circular motion), and then 2B of radius 1.5. Note that the toolpath of semi-circles 1A and 2B are nothing but transitioning into the size of the pocket specified in I3 and back to the starting position, respectively. Instead of a pocket, the cutter cuts a trajectory of two circles of depth 0.25, 1A and 2B of radius 1.5, and 1B and 2A of radius 3.

Example B: G12 I1. K3. Q1. F10. Z–.25

Again, we assume the current position of the cutter is (3, 3), as shown in Figure 2.7(b). This NC block moves the cutter by the following circular paths:

(1) semi-circle 1A of radius 0.5,
(2) semi-circle 1B of radius I = 1,
(3) semi-circle 2A of radius 1.5,
(4) semi-circle 2B of radius I + Q = 2,
(5) semi-circle 3A of radius 2.5,
(6) semi-circle 3B of radius I + 2Q = 3,
(7) semi-circle 4A of radius K = 3,
(8) semi-circle 4B of radius 1.5, which transitions the cutter back to (3, 3).

As seen in Figure 2.7(b), the cutter cuts a trajectory of four complete circles of depth 0.25, with radius varying from 0.5 to K = 3.

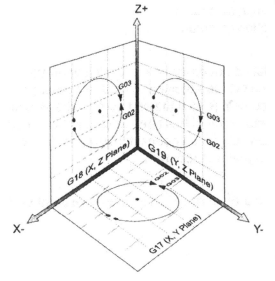

Note that if a cutter of diameter greater than 1 is employed, the NC block cuts a circular pocket. However, the outer radius of the pocket is K plus the radius of the cutter. In order to cut a pocket of radius specified by K, the cutter must be compensated by its radius. Again, we will discuss tool radius compensation in Section 2.8.

G17/18/19: XY/ZX/YZ Plane Selection

G17, G18, and G19 select XY, ZX, and YZ planes for circular motion, respectively.

Note that the direction of the circular motions is determined by viewing from positive to negative direction along the axis that is normal to the respective plane selected.

The default plane selected is XY; i.e., G17 is default.

Figure 2.8 Plane sections specified by G17, G18, and G19 commands (figure extracted from HAAS operation manual)

G81-89: Canned Cycles

A canned cycle is used to simplify programming of a part. Canned cycles are defined for mainly Z-axis repetitive operations such as drilling, tapping, and boring. Once selected a canned cycle is active until cancelled with G80. When active, the canned cycle is executed every time an X or Y-axis motion is programmed. Those X-Y motions are executed as rapid motion commands (G00) and the canned cycle operation is performed after the cutter travels to the position specified by the X and Y commands. G81 to

G89 support different hole drilling operations; for example, G81 performs spot drilling, G82 counter boring, G83 peck drilling, and so on.

There are six operations involved in every canned cycle:

- Positioning of X and Y axes
- Rapid transverse to R plane
- Drilling (spot drilling, counter boring peck drilling, and so on)
- Operation at bottom of bole
- Retraction to R plane
- Rapid transverse up to initial point

Format: G81 F_ L_ R_ X_ Y_ Z_
            L: Number of repeats
            R: Position of the R-plane
            X-Y: Center position of the hole
            Z: Position of the bottom of the hole

Example: G81 Z–1.5 F15. R.1

This NC block set up the canned cycle for spot drilling holes of 1.5 depth with feedrate 15 and retract to a 0.1 plane. Cutter is not moving until an X or Y command in the NC program is executed.

The following are default settings; that is, the control automatically reads these G words when machine power is turned on.

G00	Rapid traverse
G17	XY plane selection
G40	Cutter compensation cancel
G49	Tool length compensation cancel
G54	Work coordinate zero #1 (1 of 6 available)
G64	Exact stop cancel
G80	Canned cycle cancel
G90	Command absolute programming
G98	Canned cycle initial point return

There is no default feedrate (F code), but once an F code is programmed, it will apply until another is entered or the machine is turned off.

2.6   Machine Functions or M words

Machine functions (also referred to as miscellaneous or auxiliary functions) designate a particular mode of operation, typically to switch a machine function (such as coolant supply, spindle, and so on) on or off.

A list of M words commonly found on FANUC and similarly designed controls for milling and turning can be found in Table 2.3. Note that only one M word can be used per block. The M word will be the last action to be performed, regardless of where it is located in the block.

A complete list of M words for HAAS CNC machines can be found in Appendix E.

Table 2.3 FANUC style M words

Code:	Function:
M00	Stop Program
M02	Program End
M03	Spindle Forward
M04	Spindle Reverse
M05	Spindle Stop
M06 or 16	Tool Change
M08	Coolant On
M09	Coolant Off
M30	Program End and Rewind
M97	Local Sub-Program Call
M98	Sub Program Call
M99	Sub Program Return Or Loop

2.7   NC Part Programming

NC part programming consists of planning and documenting the sequence of processing steps to be performed on a CNC machine. The outcome is an NC part program used to instruct the machine to cut a stock (or stock material, or workpiece) to a desired part. An NC part program describes the sequence of actions of the controlled CNC machine, which include, but are not limited to, the following:

- Tool movements, including direction, velocity, and position,
- Tool selection, tool change, tool offsets, and tool compensation,
- Spindle rotation direction and spindle rotation speed,
- Cutting speed for different sequences,
- Application of cutting fluids.

In this section, we discuss machine home position, work coordinate system, template of NC part program, followed by examples.

Machine Home Position

When a CNC machine is first turned on, it does not know where the axes are positioned in the workspace. Home position is found by the Power On Restart sequence initiated by the operator who pushes a button on the machine control panel after turning on the control power. The Power On Restart sequence simply drives all three axes slowly towards their extreme limits in X, Y, and Z directions. As each axis reaches its mechanical limit, a microswitch is activated. This signals to the control that the home position for that axis is reached. Once all three axes have stopped moving, the machine is said to be "homed". Machine coordinates are thereafter in relation to this home position.

Work Coordinate System

Obviously, it would be difficult to write an NC program in relation to machine home position. The home position may be away from the jig table, so values in the NC program would be large and have no easily recognized relation to the part model.

To make programming and setting up the CNC easier, a Work Coordinate System (WCS) is established for each NC program. The WCS is a point selected by the NC programmer on the part, stock or fixture. While the WCS can be the same as the part origin in CAD, it does not have to be. Although WCS can be located anywhere in the machine envelope, its selection requires careful consideration.

The origin of a WCS is located by mechanical means such as an edge finder, coaxial indicator or part probe. It must be located with high precision: typically plus or minus 0.001 inches or less. It must be repeatable: parts must be placed in exactly the same position every time. It should take into account how the part will be rotated and moved as different sides of the part are machined. The origin of WCS is often placed at the front left corner of the top face of a rectangular stock or at the center point of the end face of a round stock.

Multiple WCS can be defined and registered to a CNC machine. For example, up to six WCS can be located on a HAAS mill. The origin of the respective WCS is called program zero, to which the NC part program refers. In the NC part program, G54 to G59 is employed to select a respective WCS. Note that it is obvious that the WCS assumed in the NC part program will have to be consistent with the WCS physically set up on the CNC machine.

Pseudo Codes (Template)

Note that in general a cutter will have to be selected first, followed by spindle and coolant on, before any cutter motion is taking place. After all the cutting motions, we turn off spindle and coolant, return the cutter to its holding position (tool turret), and select next cutter for more cutting. A template NC part program is shown below for your reference.

Line #1 = Select cutting tool (T__ M06)
Line #2 = Turn the spindle on and select the RPM (S__ M03)
Line #3 = Turn the coolant on (M08)
Line #4 = Rapid to the starting position of the part (G00)
Line #5 = Choose the proper feedrate and make the cut (F__ G__)
Line #6 = Turn off the spindle (M05)
Line #7 = Turn off the coolant (M09)
Line #8 = Return tool to holding position and select next tool (G28)

Example 2.1: Trajectory Milling

This example involves trajectory milling, in which a 0.75in. flat end mill of Tool#4 is used to cut a slot of 0.25in. deep that goes through a path connecting Points 1 to 7 and back to Point 1, as shown in Figure 2.9(a). Note that the red dot in Figure 2.9(a) is set as the program zero. The toolpath includes straight lines and circular arcs. The NC part program that cuts the slot is listed in Figure 2.9(b) with brief explanations. Note that G43 in the N25 block moves the Tool#4 to a position of Z = 0.1, in which the length of the tool has been measured and registered to the machine. G43 instructs the machine controller to extract the tool length registered and compensate the length when moving the tool to the position of Z = 0.1. More about tool length compensation G43 will be discussed in Section 2.8.

Example 2.2: Profile Milling

In this profile milling operation, Tool#1 of 0.75 in. diameter is cutting a 7 in.×4.5in.×0.25in. aluminum block along the part boundary profile with the lower left corner of the block as the program zero; see Figure 2.10(a). The cutter follows the toolpath connecting Points A to L, including straight lines and circular arc. The NC part program that performs the profile milling operation is listed in Figure 2.10(b) with a brief explanation.

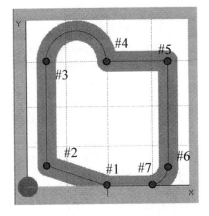

Part Program	Explanation
O0100	Program #0100
N5 G90 G40 G80 G54	Absolute programming, tool radius compensation off, WCS#1 for program zero
N10 M06 T04	Tool change to load tool #4
N15 M03 S2000	Spindle on clockwise at 2000 rpm
N20 G00 X2 Y-0.375	Rapid to X2, Y-0.375
N25 G43 Z.1 H04 M08	G43: Tool length compensation, Tool#4, M08: coolant on
N30 G01 Z-0.25 F15	Feed down to Z-0.25 at 15 ipm
N35 Y0	Feed move to point #1
N40 X.5 Y.5	Feed move to point #2
N45 Y3.0	Feed move to point #3
N50 G02 X2 I0.75	Circular feed move to point #4
N55 G01 X3.5	Feed move to point #5
N60 Y.5	Feed move to point #6
N65 G02 X3 Y0 I-0.5	Circular feed move to point #7
N70 G01 X2	Feed move to point #1
N75 G00 Z1	Rapid to Z1
N80 X0 M09	Rapid to X0, coolant off
N85 M05	Spindle off
N90 M30	End of program

(a)   (b)

Figure 2.9 Example 2.1: Trajectory Milling, (a) sketch of the toolpath, and (b) the NC program with brief explanations

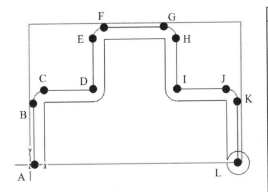

Part Program	Explanation
O0100	Program #0100
N0010 T1 M06	M06: Tool change, Load Tool#1
S580 M03	M03: Spindle forward
G00 X.125 Y0.	G00: Rapid Motion, Move cutter to X.125Y0 (Point A)
G43 Z.1 H01 M08	G43: Tool length compensation, Tool#1, M08: coolant on
G01 Z-.25 F2.3	G01: Linear interpolation, Z=-.25, Cutting, feed rate 2.3 in/min
Y2.	Linear interpolation, Move to Y2.0 (Point B)
G02 X.5 Y2.375 I.375 J0.	G02: CW interpolation, Center= (0.5,2.375) (Point C)
G01 X2.125	G01: Linear interpolation, X2.125Y2.375 (Point D)
Y4.	(Point E)
G02 X2.5 Y4.375 I.375 J0.	(Point F)
G01 X4.5	(Point G)
G02 X4.875 Y4. I0. J-.375	(Point H)
G01 Y2.375	(Point I)
X6.5	(Point J)
G02 X6.875 Y2. I0. J-.375	(Point K)
G01 Y0.	(Point L)
Z1.	Retract to Z = 1
M05	Spindle off
M30	End of program

(a)   (b)

Figure 2.10 Example 2.2: Profile Milling, (a) sketch of the toolpath, and (b) NC program with brief explanations

Subroutines

Subroutines allow the CNC programmers to define a series of commands that might be repeated several times in a program and, instead of repeating them, they can be "called". A subroutine call is made with M97 or M98 and a P command.

Example 2.3: Hole Drilling

This example involves canned cycle calls G81 for hole drilling, and a subroutine call using M98 for hole locations. The NC part program performs a center drill operation for 9 holes arranged in three rows evenly spaced—see Figure 2.14(a)—using a ½-in. center drill (Tool#1). The operation is followed by a hole drilling operation using a ½-in drill bit (Tool#2). The X- and Y-locations of the hole centers are programmed in a subroutine O0200, as listed in Figure 2.14(b). The drill canned cycle G81 is called twice. First call is for center drill using Tool#1 that plunges into the stock with a small depth of 0.1in. for the nine holes. The second G81 call is for hole drilling operation using Tool#2 for a hole depth of 0.5in. A subroutine call is made with M98 and a P command followed by the subroutine number, in this case, 0200.

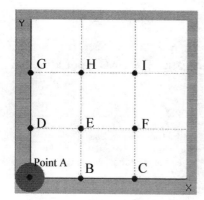

Part Program	Explanation
O0100	Main program, program #0100
T01 M06	Load Tool#1 (for center drill)
G54 G00 G90 X0. Y0.	G54: Select work coordinate system (default), G90: Absolute (default)
G43 H01 Z.1	G43: Tool length compensation
G81 R0.1 Z-0.1 F20. L0	No operation, just define drill canned cycle, G81: drill canned cycle*
S2000 M03	M03: Spindle forward
M98 P0200	M98: Subroutine call, P0200: call subroutine program #0200, center drill each hole (total: 9 holes)
M05	M05: Spindle off
T02 M06	Load Tool#2 (for drill)
G81 R0.1 Z-0.5 F30. L0	No operation, just define drill canned cycle, G81: drill canned cycle*
S3000 M03	M03: Spindle forward
M98 P0200	M98: Subroutine call, P0200: call subroutine program #0200, drill each hole (total: 9 holes 3×3)
M05	M05: Spindle off
G28	G28: Return to reference point
M30	M30: End of main program
O0200	Subroutine program #0200, listing all holes
X0. Y0.	Point A
X1.	Point B
X2.	Point C
X0. Y1.	Point D
X1.	Point E
X2.	Point F
X0. Y2.	Point G
X1.	Point H
X2.	Point I
M99	End of subroutine

(a)   (b)

Figure 2.11 Example 2.3: Hole drilling, (a) sketch of the toolpath, and (b) NC program with brief explanations

2.8   Tool Functions and Compensations

Selecting tools and setting up accurate compensations (including tool length and tool radius compensations) are important steps in NC part programming. In particular, pocket milling (for example G12/13 discussed in Section 2.5) or profile milling that cuts geometric features to their precise dimensions often requires turning on tool radius compensations so that the program can be reused for different size cutters.

In this section, we discuss tool change, tool length compensation, and tool radius compensations, followed by examples for illustration.

Tool Functions (Tnn)

The Tnn code is used to select a tool to be placed in the spindle from the tool changer. The T address does not start the tool change operation; it only selects which tool will be used next. M06 and M16 are used to start a tool change operation. The T word can be in the same block as the M06 or M16.

Tool Length Compensation: G43

G43 specifies tool length compensation in a positive direction. That is, the tool length offsets are added to the commanded axis positions (usually Z-axis). A nonzero Hnn must be programmed to select the correct entry from the length offset registry.

For example, the NC block G43 H01 Z.1 instructs the machine controller to move Tool#1 (H01) tool to Z = 0.1 position, as illustrated in Figure 2.12.

At the shop floor, length of Tool#1 in the tool turret must be measured beforehand and registered to the machine memory.

G49 cancels tool length compensation.

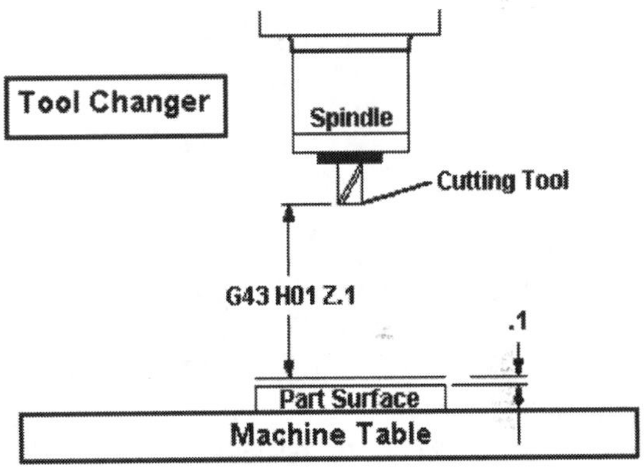

Figure 2.12 Illustration of tool length compensation G43 (figure extracted from HAAS operation manual)

Cutter (Tool) Radius Compensation; G41/42

G41/42 commands turn on tool radius compensation, which are modal commands. The commands position the tool further away from the desired part or pocket wall in order to accommodate the radius of the cutting tool.

The desired boundary of the part, either pocket or part profile boundary, to cut is referred to as programmed path, and the toolpath is compensated away from the programmed path by the amount of the tool radius. The NC part program is written by entering X and Y positions that follow the programmed path when G41 or G42 is taking place.

G41 selects cutter compensation left; that is the tool is moved to the left of the programmed path to compensate for the radius of the tool, as illustrated in Figure 2.13(a). A D command must also be programmed to select the correct tool size from those registered to machine memory.

G42 moves the tool to the right of the programmed path, as shown in Figure 2.13(b). The left or right of the programmed path is determined by the tool-motion direction. In general, CNC mills read multiple blocks of code and store them in machine buffer. Left or right of the programmed path can therefore be determined. For example, a HAAS mill reads the next 16 blocks and keeps them in buffer. The machine controller does know "when to stop" and "where to turn" a couple of steps back.

G41/42 simplifies the NC part programming significantly. This is because figuring out the X and Y locations that define the boundary profile of a part or a pocket; i.e., the programmed path, is much easier than those of the toolpath that incorporates the offset of the size of the cutting tool. Moreover, the NC part program becomes more flexible while employing tool radius compensation. This is because the part program is not associated with a cutter of prescribed size. Different size tools produce the same outcome as long as size registration is correctly selected in the program using a D command.

Note that G40 cancels tool-radius compensation. G41 and G42 are modal and are applicable only for G17; i.e., when cutting on the X-Y plane is selected.

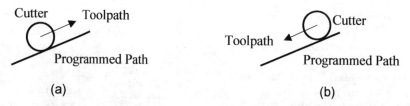

Figure 2.13 Illustration of tool radius compensations, (a) G41 compensation from the left, and (b) G42 compensation from the right

Example 2.4: Trajectory Milling and Profile Milling with Tool Radius Compensation

This is a continuation of Example 2.2; see sketch of the part in Figure 2.14(a). This example involves two operations, trajectory milling and profile milling. A small cutter of diameter 0.25in. (Tool#1) is loaded first to cut a 1/8in.-deep trajectory of straight lines that connect Points P1 to P8 on a 7in.×4.5in.×0.25in. aluminum block, as illustrated in Figure 2.14(b). A larger cutter of 0.75in. flat end mill (Tool#4) is employed for the profile milling operation, in which tool radius compensation left (G41) is turned on. The resulting toolpath includes straight lines and circular arcs that turn the cutter around the corners. The NC part program that performs both operations is listed in Figure 2.14(c) with brief explanations. Note that in this example, Point P0 at the lower left corner of the stock is defined as the program zero.

Comparing this example with that of Example 2.2, the advantage of tool radius compensation for manual NC part programming is clear.

Example 2.5: Pocket Milling with Tool Radius Compensation

In this example, we cut a circular pocket of radius 3in. and 0.25in. deep using a cutter of 1in. flat end mill. The lower left corner of the 6in.×6in.×1in. stock is chosen as the program zero. The cutter first transverses to the center of the pocket at (3, 3), as shown in Figure 2.15(a). Then, G42 is turned on for tool radius compensation from the right. Note that D01 is included in the block that contains G42. D01 instructs the machine controller to extract radius information of Tool#1 registered for accurate radius compensation. A clockwise pocket milling command G12 is then called to perform the pocket milling. The programmed path of the pocket milling shown in Figure 2.15(b) is identical to that of Figure 2.7(b), and the toolpath is illustrated in Figure 2.15(c).

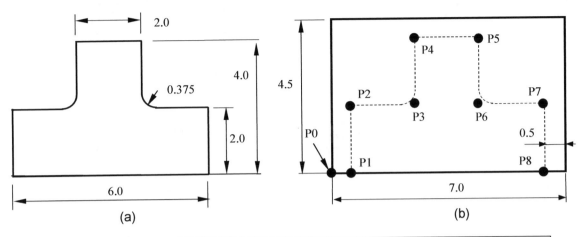

Part Program	Explanation
O0100	Program #0100
N5 M06 T01	Tool change to Tool #1
N10 G54 G00 G90 X0.5 Y0. Z1.	Set default, and move cutter to Point P1: X0.5 Y0. Z1.
N15 G43 H01 Z.1	G43: tool length compensation, cutter positioned to Z=0.1
N20 S580 M03	M03: Spindle forward
N25 G01 Z-0.125 F2.3	G01: cutting, feed rate 2.3 in/min
N30 Y2	Feed move to point P2
N35 X2.5	Feed move to point P3
N40 Y4	Feed move to point P4
N45 X4.5	Feed move to point P5
N50 Y2	Feed move to point P6
N55 X6.5	Feed move to point P7
N60 Y0	Feed move to point P8
N65 G00 Z1.	Rapid retract
N70 X0 Y0 M05	Rapid to X0 Y0, spindle off
N75 M06 T4	Tool change to Tool #4
N80 M03 S2000	Spindle on clockwise at 2000 rpm
N85 G41 D04 X0.5 Y0	**Compensation on, rapid move to point P1**
N90 G01 Z-0.25 F15 M08	Feed down to Z-0.25 at 15 ipm, coolant on
N95 Y2	Feed move to point P2
N100 X2.5	Feed move to point P3
N105 Y4	Feed move to point P4
N110 X4.5	Feed move to point P5
N115 Y2	Feed move to point P6
N120 X6.5	Feed move to point P7
N125 Y0	Feed move to point P8
N130 G00 Z1	Rapid retract
N135 G40 X0 Y0 M09	Rapid to X0 Y0, radius compensation cancelled, coolant off
N140 M05	Spindle off
N145 M30	End of program

(c)

Figure 2.14 Example 2.4: Trajectory milling and profile milling with tool radius compensation, (a) sketch of the part, (b) part overlapping with the stock, and (c) NC program with brief explanations

Part Program	Explanation
O0100	
T1 M06	Tool 1 is a 1 inch diameter end mill
G54 G00 G90 X0. Y0.	
G43 H01 Z.1	:Tool length compensation
S2000 M03	
G00 X3. Y3.	
G42 G00 D01 X3. Y3.	**Set tool radius compensation**
G12 I1 K3 Q1 F10. Z–.25	Outer pocket diameter 6
G00 Z.1	
G40	Cancel tool radius compensation
G28 X0. Y0.	
M30	

(a)

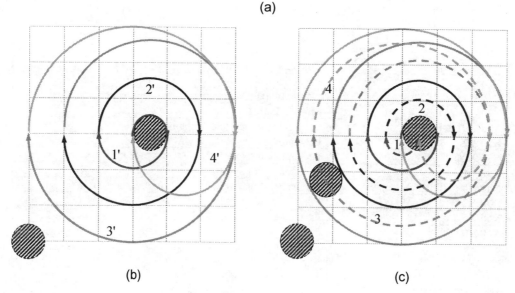

(b)   (c)

Figure 2.15 Example 2.5 : Pocket milling with tool radius compensation, (a) NC program with brief explanations, (b) and sketch of the programmed path, and (c) sketch of the toolpath in dotted lines

2.9   NC Program Review and Verification

Although simple and straightforward, as mentioned before, manual programming is not necessarily the best option in generating NC part programs. First, it is often time consuming to figuring out cutter locations based on part geometry and cutter size. This is especially true while cutting a freeform (or sculptural) surface. Usually, a few predefined programs (canned cycles, such as hole-drilling) come with the CNC unit, which makes manual part programming easier. Second, the manual approach is limited in supporting simple machining work (e.g., profile milling), in which cutter locations can be easily acquired. Third, although syntax and semantics of the NC program are fairly standardized, differences do exist between machine manufacturers (or the controller employed). As a result, an NC part program that works on one machine may not work on another.

It is critical that NC programmers carefully review and verify the NC programs before uploading them to CNC machines to cut parts. This can be done by either using a NC reviewer, such as SOLIDWORKS

CAM NC Editor—see Figure 2.16(a)—to be discussed in Lesson 3, or directly on the monitor screen of the CNC machine; e.g., HAAS 3-axis mill shown in Figure 2.16(b).

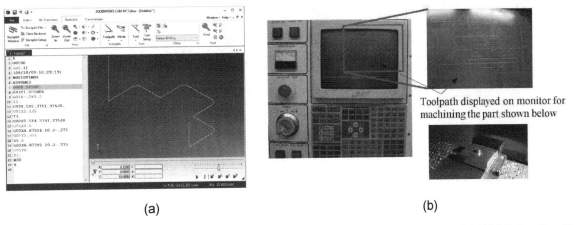

(a)   (b)

Figure 2.16 NC code verification using (a) SOLIDWORKS CAM NC Editor, and (b) HAAS 3-axis mill (www.haas.com)

2.10 Exercises

Problem 2.1. Write an NC program (with explanations for each NC block) for a HAAS mill to machine a 4in.×4.25in.×1.55in. aluminum block for the part (design model) shown in Figure 2.17(a). Choose your own cutters, from those shown in Figure 2.17(b) and cutting parameters (step depth and step over) with "common sense". Note that you need to use tool radius compensation for all toolpath. Calculate the overall cutting time.

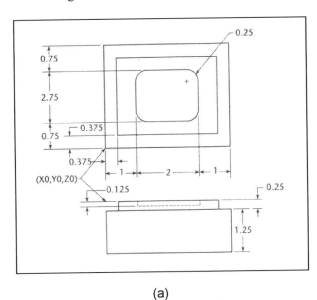

Size (inches)	Type	Length (inches)
1/8	Ball-nose	0.375
3/16	Ball-nose	1.000
1/4	Ball-nose	0.625
1/2	Ball-nose	2.000
1/8	Flat-end	0.375
3/16	Flat-end	0.475
1/4	Flat-end	0.500
1/4	Flat-end	0.875
5/16	Flat-end	0.600
3/8	Flat-end	0.600
1/2	Flat-end	1.000
1/2	Flat-end	1.500

(a)   (b)

Figure 2.17 Part and tools for Problem 2.1

Problem 2.2. Write an NC program to cut a 0.25in.-deep pocket on a 3in.×3in.×0.5in. block shown in Figure 2.18. The cutter diameter is 0.5 in. Note that the cutter must cut inside of the pocket by moving along the pocket boundary only one time, which will leave some material uncut. You must use tool radius compensation for composing the NC program. Identify the uncut area. Also, sketch your toolpath, which must start from Point A (0,0).

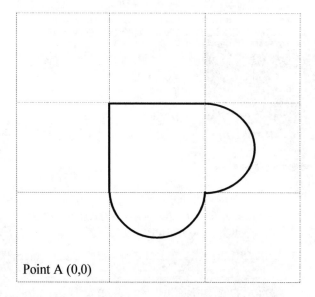

Point A (0,0)

Figure 2.18 The shape of the pocket of Problem 2.2

Lesson 3: SOLIDWORKS CAM NC Editor

3.1 Overview of the Lesson

As discussed in Lesson 2, NC part programs manually composed will have to be verified before uploading them to a CNC machine for part machining. In this lesson, we focus on learning to use SOLIDWORKS CAM NC Editor, or simply NC Editor, for reviewing and verifying NC programs.

NC Editor is a software module embedded in SOLIDWORKS CAM, which facilitates users to compose new NC part programs or edit an existing G-code by reviewing and visualizing the corresponding toolpath in a graphical window. Users may simply copy and paste a few blocks of an NC program and visualize the toolpath defined by the blocks for review and verification. Also, users may select cutters as desired from the tool library, create new tools or modify existing tools, and turn on tool radius compensation to review the toolpath of the associative NC program. In addition, NC Editor offers useful capabilities, such as an enhanced file compare utility that supports users to compare NC programs for merging or further editing.

We will go over a short overview of NC Editor, and use four examples, both milling and turning operations, to help you become familiar with the capabilities the module offers. In addition, we will briefly mention a few good G-code viewers and editors on-line that may be of interest to you.

After completing this lesson, you should be able to use NC Editor to review, edit, and verify NC programs and become aware of other similar tools on-line.

3.2 Capabilities of SOLIDWORKS CAM NC Editor

SOLIDWORKS CAM NC Editor offers two major capabilities, editor and backplot. The editor, which is part of NC Editor, offers a set of editing capabilities necessary for NC program editing. It has no program size limitations and includes NC program specific options such as line numbering/renumbering, character handling and XYZ range finder. It also features mathematic functions, including basic math, rotate, mirror, tool compensation, and translate. In addition, the editor offers common functions expected from an editor including drag-and-drop text editing.

The backplot supports visualization of 3-axis mill and 2-axis lathe NC part programs with step and continuous forward and reverse plotting. When users edit the NC program, the toolpath is updated and is automatically reflected in the plot. Users may analyze the plot with dynamic zoom, pan, rotate and measuring functions. The backplot also supports visualization of NC programs with tool holder collision check and gouge detection.

In addition, NC Editor features a side-by-side file compare, allowing the user to quickly identify changes made in an NC program. The file compare identifies changed and deleted/inserted lines but ignores trivial format changes like block renumbering and spacing. Differences are displayed one line at a time, all at once or printed side-by-side for offline review.

3.3 User Interface

You may start NC Editor from within SOLIDWORKS CAM. From SOLIDWORKS CAM, you may first open a solid model, choose SOLIDWORKS CAM tab as circled in Figure 3.1, and click *SOLIDWORKS CAM NC Editor* button (also circled in Figure 3.1). NC Editor will appear like that of Figure 3.2 (first without any cutter, toolpath, or NC program).

Figure 3.1 Starting SOLIDWORKS CAM NC Editor from within SOLIDWORKS CAM

NC Editor has a standard Windows® user interface (see Figure 3.2), having one or more windows for every open file. The functions in NC Editor can be activated through the command buttons of the *Ribbon Bar*, which help you to quickly find the commands needed for a task. The command buttons are organized into logical groups under individual functions tabs, including *Editor*, *NC Function*, *Backplot*, and *Transmission*. The command buttons under the respective command tabs offer the functions required to compose, modify, and verify the NC program.

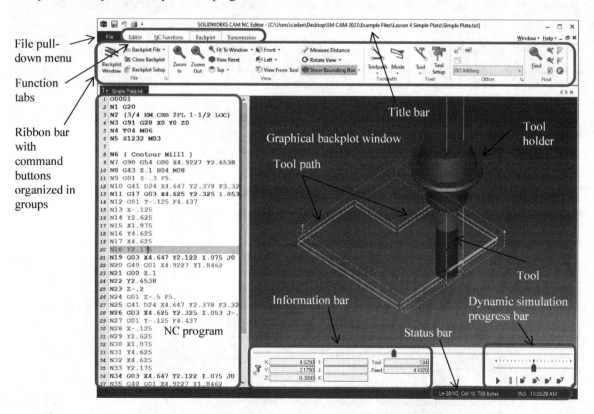

Figure 3.2 User interface of SOLIDWORKS CAM NC Editor

The *Editor* tab allows users to create new NC programs or open/close an existing NC program, as well as set up global selections for the entire program .

Commands in the *NC Functions* tab support users to build or change NC programs.

The *Backplot* tab of NC Editor offers capabilities to support the simulation of NC programs. Command buttons under the *Backplot* tab are used for the setup and simulation of NC programs. The backplot window shows the toolpath of milling and turning operations. In *Backplot* mode, the NC program is displayed in the left pane of the window, while the plot shown in the right pane displays the toolpath of the NC program and tool. You can stop and restart the simulation, control the speed and direction of the tool movement, etc. During the dynamic simulation, a gray bar appears in the NC program, indicating the specific NC block being processed. You can simulate an NC program for milling operations using wireframe or solid view. For turning operations, you can only use wireframe simulation.

In addition, the *Backplot Information Bar* shown in the lower part of the *Backplot* window displays the current cutter location within the defined preparatory functions. The slider of the *Dynamic Simulation Progress Bar* allows users to alter the progress of the simulation.

The *Transmission* tab allows users to send and receive NC programs to their CNC machines at the shop floor.

To the left of the function tabs is the *File* pull-down menu, which provides basic file open/close/print functions.

3.4   NC Program Examples

Four NC program examples, including *Example1_Trajectory_Milling.txt*, *Example2_Hole_Drilling.txt*, *Example3_Profile_Milling.txt*, and *Example4_Turning_Finish.txt*, are prepared for you to browse the capabilities and become familiar with the use of NC Editor. Example 1 performs a trajectory milling that cuts a slot of 0.75in. wide and 0.25in. deep. Example 2 includes canned cycle calls for hole drilling operations and a subroutine call. Example 3 involves profile milling operation with tool radius compensation. Example 4 presents a turning operation. Again, these files are available for download at the publisher's website (www.sdcpublications.com). You may review the following sections for steps to bring the examples into NC Editor.

3.5   Example 1: Trajectory Milling

The same example discussed in Lesson 2 is repeated to illustrate the use of NC Editor. This example involves trajectory milling, in which a 0.75in. flat end mill of Tool#4 is used to cut a slot of 0.25in. deep that goes through a path connecting Points 1 to 7 and back to Point 1, as shown in Figure 3.3(a). The toolpath includes straight lines and circular arcs. The NC program that cuts the slot is listed in Figure 3.3(b) with brief explanations. The text file, *Example1_Trajectory_Milling.txt*, can be found in the Lesson 3 folder of the example files. In this example, we learn how to bring an NC program into NC Editor, choose cutter, and run toolpath using the backplot window.

Bringing in the NC Program

Select the *Editor* tab at the top of the NC Editor—circled in Figure 3.4(a)—and choose *Open*. Choose the folder where the NC program, *Example1_Trajectory_Milling.txt*, is located, and then open the file. The NC program will appear; see Figure 3.4(b).

Note that *ISO Milling* [circled in Figure 3.4(a)], which is a generic CNC mill, is selected as the machine type by default. Click the *Machine Templates* button ▣ (the fourth button from the left, circled in Figure 3.4(a), which is right underneath *ISO Milling*) to bring up the machine template in the *Setup Machine* dialog box, as shown in Figure 3.5.

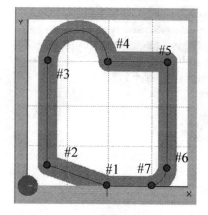

Part Program	Explanation
O0100	Program #0100
N5 G90 G40 G80 G54	Absolute programming, tool radius compensation off, WCS#1 for program zero
N10 M06 T04	Tool change to load tool #4
N15 M03 S2000	Spindle on clockwise at 2000 rpm
N20 G00 X2 Y-0.375	Rapid to X2, Y-0.375
N25 G43 Z1. H04 M08	G43: Tool length compensation, Tool#4, M08: coolant on
N30 G01 Z-0.25 F15	Feed down to Z-0.25 at 15 ipm
N35 Y0	Feed move to point #1
N40 X.5 Y.5	Feed move to point #2
N45 Y3.0	Feed move to point #3
N50 G02 X2 I0.75	Circular feed move to point #4
N55 G01 X3.5	Feed move to point #5
N60 Y.5	Feed move to point #6
N65 G02 X3 Y0 I-0.5	Circular feed move to point #7
N70 G01 X2	Feed move to point #1
N75 G00 Z1	Rapid to Z1
N80 X0 M09	Rapid to X0, coolant off
N85 M05	Spindle off
N90 M30	End of program

(a)   (b)

Figure 3.3 Example 1: Trajectory Milling, (a) sketch of the toolpath, and (b) the NC program with brief explanations

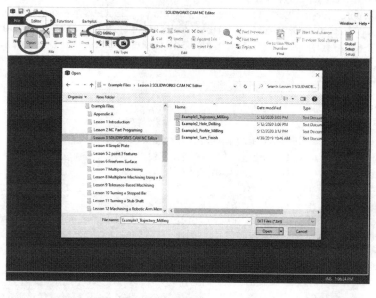

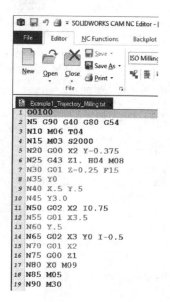

(a)   (b)

Figure 3.4 Bringing NC program into SOLIDWORKS CAM NC Editor, (a) open the NC program file, and (b) NC program appearing in the NC Editor

The basic machine information can be found; for example, *inches* is chosen for *Toolpath unit* (circled in Figure 3.5). You may modify the machine template; for example, choose *inch/min* for *Feedrate unit* if it has not been chosen. Click *OK* to accept the change and close the dialog box. We will use the *ISO Milling* for this example.

The Backplot Tab

Choose *Backplot* tab and click the *Backplot Window* button ≋ (the first button from the left, circled in Figure 3.6). The NC Editor window splits into two. The left pane lists the NC program, and the right pane shows a toolpath with no cutter initially.

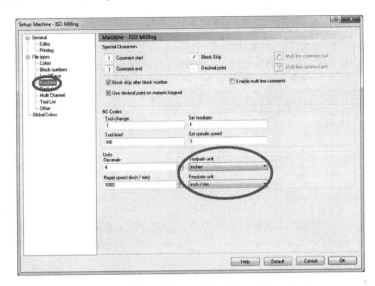

Figure 3.5 The *Setups Machine* dialog box

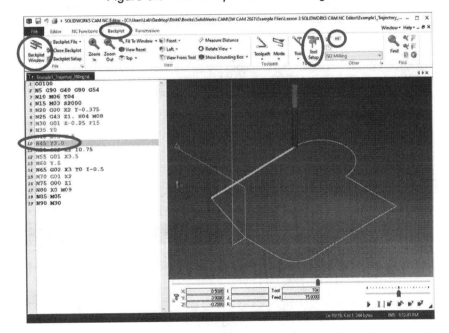

Figure 3.6 The command buttons in *Backplot* tab and toolpath displayed in the backplot window

If you click an NC block, for example, *N45 Y3.0*, the corresponding tool movement is highlighted in thick yellow color in the backplot window, as shown in Figure 3.6. Note that the default cutter of 1/8" appears, which is not the one with correct diameter of 0.75in. We will select a correct cutter next.

Tool Setup

Choose *Tool Setup* button (circled in Figure 3.6) to bring up the *Tool Setup* dialog box [see Figure 3.7(a)]. In this dialog box you can load and save tool libraries; add, edit, or delete existing tools in the library. Tool libraries are especially useful on machines with fixed tool position. You can also select a tool in the tool list of the bottom half of the window with one click. You may double-click it to get the tool dimensions. You can change one or more dimensions of the tool, enter a tool name and then assign it to the NC program.

A default tool, *End Mill Flat 1/8"*, has been chosen. This is why we see a 1/8" end mill in Figure 3.6.

In this example, we choose *End Mill Flat* for *Type* [circled in Figure 3.7(b)], select *End Mill Flat ¾"*, choose *Assign to 'T04'*, and click *OK*. Note that in the NC program, *N10 M06 T04* loads Tool#4, hence we assign *End Mill Flat ¾"* to T04. The 3/4in. cutter now replaces the default cutter in the backplot window as seen in Figure 3.8.

Work Coordinate System (WCS)

The NC word G54 in the first block (N5) selects the first work coordinate system (out of six) as the program zero for the NC program. You may need to make sure the program zero is properly selected for NC Editor.

To select the program zero for G54, we choose the *Set Workpiece/Tool Offsets* button (circled in Figure 3.6) of the *Other* group to bring up the *Workpiece/Tool Offsets* dialog box (see Figure 3.9). The offsets X, Y, and Z are all set to be zero, which is desirable since the program zero is assumed at (0,0,0) of a coordinate system setup beforehand at the CNC machine. Click *Cancel* to close the dialog box.

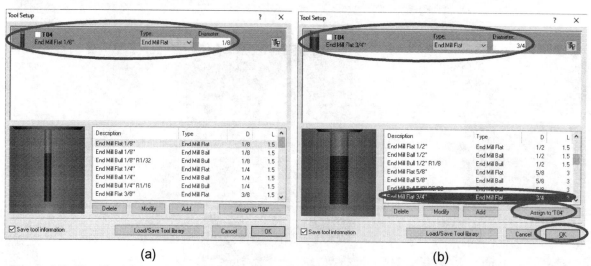

(a)   (b)

Figure 3.7 The *Tool Setup* dialog box, (a) End Mill Flat 1/8" selected as the default tool, and (b) End Mill Flat 3/4" assigned as the default tool

Simulating the Toolpath

You may click the *Start/stop simulation* button ▶ in the *Dynamic Simulation Progress Bar* (circled in Figure 3.8) to review the toolpath.

More Options for Toolpath Simulation

You may pull down the *Toolpath* button _{Toolpath} to select more options, as shown in Figure 3.10(a). For example, selecting *Show Points* will show CL data of the toolpath in the backplot window. You may choose numerous simulation modes by pulling down the *Mode* button _{Mode}; see Figure 3.10(b). For example, selecting *Loop Simulation* will keep the simulation running and repeating itself continuously.

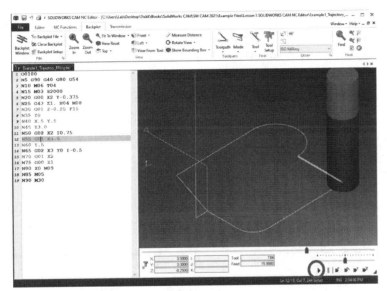

Figure 3.8 The *Backplot Window* showing ¾" end mill

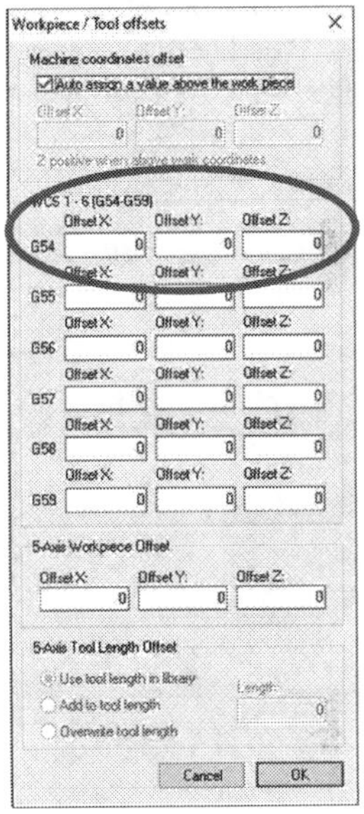

Figure 3.9 The *Workpiece/Tool Offsets* dialog box

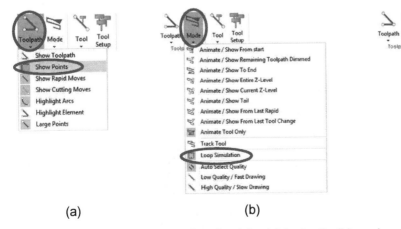

(a)   (b)   (c)

Figure 3.10 Selections available for function tabs (a) toolpath, (b) mode, and (c) tool to adjust display in the backplot window

You may also pull down the *Tool* button ![tool] to select more options, as shown in Figure 3.10(c). For example, selecting *Show Tool Holder* will display a tool holder like that of Figure 3.6 in the backplot window.

Click the *Editor* tab and select *Close* to close the NC program. We have finished the first example.

3.6   Example 2: Hole Drilling

The second NC program is identical to that of Example 2.3 discussed in Lesson 2. The NC program involves a canned cycle call of G81 for hole drilling, and a subroutine call using M98. The NC program performs a center drill operation for 9 holes in three rows evenly spaced—see Figure 3.11(a)—using Tool#1 (1/4" center drill). The operation is followed by a hole drilling operation using Tool#2 (1/2" drill bit). The X- and Y-locations of the hole centers are programmed in a subroutine O0200, as listed in Figure 3.11(b). The drill canned cycle G81 is called twice. First call is for center drill using Tool#1 that plunges into the stock with a small depth of 0.1in. for the nine holes. The second G81 call is for hole drilling operation using Tool#2 for a hole depth of 0.5in. A subroutine call is made with M98 and a P word followed by the subroutine number (in this case, 0200).

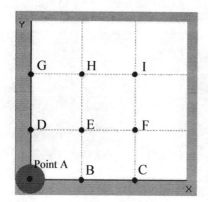

Part Program	Explanation
O0100	Main program, program #0100
T01 M06	Load Tool#1 (for example, for center drill)
G54 G00 G90 X0. Y0.	G54: Select work coordinate system (default), G90: Absolute (default)
G43 H01 Z.1	G43: Tool length compensation
G81 R0.1 Z-0.1 F20. L0	No operation, just define drill canned cycle, G81: drill canned cycle*
S2000 M03	M03: Spindle forward
M98 P0200	M98: Subroutine call, P0200: call subroutine program #0200, center drill each hole (total: 9 holes)
M05	M05: Spindle off
T02 M06	Load Tool#2 (for drill)
G43 H02 Z.1	G43: Tool length compensation
G81 R0.1 Z-0.5 F30. L0	No operation, just define drill canned cycle, G81: drill canned cycle
S3000 M03	M03: Spindle forward
M98 P0200	M98: Subroutine call, P0200: call subroutine program #0200, drill each hole (total: 9 holes 3×3)
M05	M05: Spindle off
G28 M30	G28: Return to reference point (machine zero position), M30: Program end and rewind, End of main program
O0200	Subroutine program #0200, listing all holes
X0. Y0.	Point A
X1. Y0.	Point B
X2. Y0.	Point C
X0. Y1.	Point D
X1. Y1.	Point E
X2. Y1.	Point F
X0. Y2.	Point G
X1. Y2.	Point H
X2. Y2.	Point I
M99	End of subroutine

(a)   (b)

Figure 3.11 Example 2: Hole drilling, (a) sketch of the toolpath, and (b) the NC program with brief explanations

In this example, we assume a HAAS mill. The NC program written for a HAAS mill is listed in Figure 3.11(b).

Choosing HAAS Milling and Backplot

We first click the *Editor* tab and choose *Haas Milling* as the machine type. Like that of Example 1, you may bring up the machine template by clicking the *Machine Templates* button ✉ (the fourth button from the left, right underneath *Haas Milling*). The basic machine information can be found. Like those of Example 1, we choose *inch*es for *Toolpath unit* and *inch/min* for *Feedrate unit*.

We follow the same steps as those of Example 1 to bring in the NC program (*Example2_Hole_Drilling*) for this example, and start the backplot window. Note that you will need to select *TXT Files (*.txt)* in the *Open* dialog box to see the NC program files with file suffix *.txt*.

Choose *Backplot* tab and click the *Backplot Window* button ⇛. The NC Editor window splits into two. The left pane lists the NC program, and the right pane shows a toolpath with a default cutter like that of Figure 3.12.

Tool Setup

Choose *Tool Setup* button ⚒ (circled in Figure 3.6) to bring up the *Tool Setup* dialog box [see Figure 3.13(a)]. In this example, we choose *Center Drill 1/2"* and assign it to Tool#1 (by clicking *Assign to 'T01'*). We then click the *T02* row at top of the dialog box [see Figure 3.13(b)], select *Drill 1/2"* and assign it to Tool#2 (by clicking *Assign to 'T02'*). We click *OK* to accept the tool setup.

By default, G54 picks the origin of the coordinate system set up at the CNC mill as the program zero. This is because in the *Workpiece/Tool Offsets* dialog box, offsets in all three axes are set to be zero by default (see Figure 3.9). Therefore, there is no need to open the dialog box to change anything.

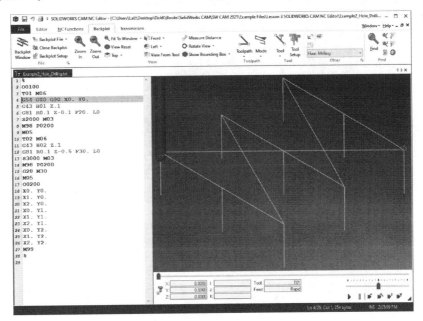

Figure 3.12 The NC program and toolpath

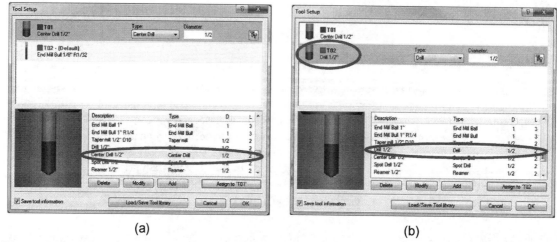

(a)   (b)

Figure 3.13 Choosing tools in the *Tool Setup* dialog box, (a) T1: Center Drill ¼", and (b) T02: Drill ½"

Drill Canned Cycle G81

As mentioned earlier, the drill canned cycle G81 is called twice in the NC program. Recall that the G81 NC word follows the format below:

G81 F_ L_ R_ X_ Y_ Z_

where L, R, X, Y, and Z are the number of repeats, position of the retract plane, center position of the hole (X and Y), and position of the bottom of the hole (Z), respectively. In this NC program, the first G81 call is G81 R0.1 Z–0.1 F20. L0, which sets up a center drill operation for a depth of 0.1in. at 20 in./min feedrate and retracts to a 0.1in. plane above the Z-axis after each center drill. Note that the repeat L is set to be zero. No action is taken when the controller of a CNC mill (or in this case the NC Editor) executes this command. The cutter will act when the first X or Y command is encountered, in this case, when the subroutine is called. Similarly, the second G81 call, G81 R0.1 Z–0.5 F30. L0, drills holes of 0.5in. deep at the same nine locations specified in the subroutine.

Simulating the Toolpath

You may click the *Start/stop simulation* button ▶ in the *Dynamic Simulation Progress Bar* to review the toolpath like that of Figure 3.14.

The toolpath shows a sequence of center drill of the first G81 call by using tool T01 and cutting holes of depth 0.1in. at points A, B, and C in the first row; then the cutter moves to point D, and cuts the holes of the second row at points D, E, and F. Similarly, the cutter cuts holes in the third row (G, H, and I). The cutter retracts to Z=0.1in. after each center drill. The program then loads T02 (T02 M06), moves the cutter back to point A at 0.1in. above Z-axis, and follows the same order to drill the nine holes of 0.5in. depth, retracting to Z = 0.1in. after each drill.

Click the *Editor* tab and select *Close* to close the NC program. We have finished the second example.

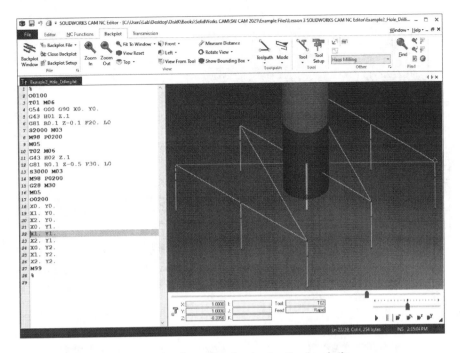

Figure 3.14 Looping the toolpath simulation

3.7   Example 3: Profile Milling

We review a profile milling example similar to Example 2.4 discussed in Lesson 2, in which a cutter of 0.75in. diameter is employed to cut a 7in.×4.5in.×0.25in. aluminum block—see Figure 3.15(b)—along the boundary profile of the design model; see Figure 3.15(a).

The NC program is listed in Figure 3.16 with brief explanations. Note that the tool radius compensation from the left, G41, is turned on (see block N30), and the program is written to machine the part boundary by following the program path from point P1 to P8.

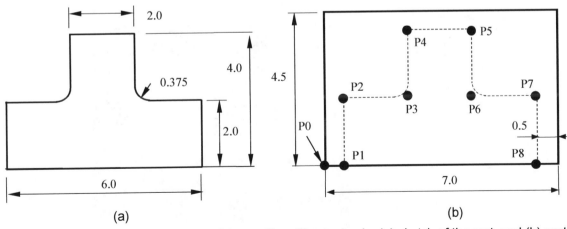

(a)   (b)

Figure 3.15 Geometric dimensions of the profile milling example, (a) sketch of the part, and (b) part overlapping with the stock

N5 T01 M06	Tool change to Tool #1
N10 G54 G00 G90 X0. Y0. Z1.	Set default and move cutter to X0. Y0. Z1.
N15 S580 M03	M03: Spindle forward
N20 G43 H01 Z.1	G43: tool length compensation, cutter positioned to Z=0.1
N25 G01 Z-0.25 F2.3 M08	G01: cutting, feed rate 2.3 in/min, M08: coolant on
N30 G41 D1	**Compensation on**
N35 X.5 Y0.	Feed move to point P1
N40 Y2.0	Feed move to point P2
N45 X2.5	Feed move to point P3
N50 Y4.0	Feed move to point P4
N55 X4.5	Feed move to point P5
N60 Y2.0	Feed move to point P6
N65 X6.5	Feed move to point P7
N70 Y0.	Feed move to point P8
N75 G00 Z1.	Rapid retract
N80 X0 Y0 M05	Rapid to X0 Y0, spindle off
N85 M09	Coolant off
N90 M30	End of program

Figure 3.16 Example 3: Profile milling, NC program with explanations

Bringing in the NC Program

Select the *Editor* tab at the top of the NC Editor, choose *ISO Milling*, and choose *Open*. Select the file folder where the NC program, *Example3_Profile_Milling.txt*, is located, and then open the file. The NC program is now listed. Choose *Backplot* tab, and click the *Backplot Window* button ⩘. The toolpath with a default cutter appears like those of Figure 3.17.

Click an NC block, for example line 11 (N55 X4.5), to move the cutter to point P5, where X: 4.5, Y: 4.0. Take a closer look at the locations of the cutter shown in the *Information Bar* below the backplot window, which indicate X: 4.5, Y: 4.0, and Z: –0.25. Note that the toolpath shown in the backplot window does not include radius compensation. Also, a default cutter selected must be replaced by the desired end mill of 0.75in.

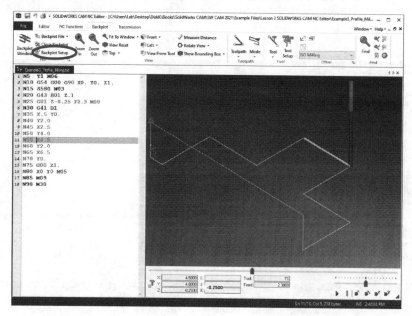

Figure 3.17 Example 3: Profile milling, the NC toolpath without radius compensation

We first follow the same steps as discussed before to select the desired cutter: end mill of 0.75in. Next, we learn to turn on tool radius compensation in NC Editor to review the toolpath with tool radius compensation.

Tool Radius Compensation

Under the *Backplot* tab, choose *Backplot Setup* button ![Backplot Setup icon] of the *File* group (circled in Figure 3.17) to bring up the *Setups Backplot* dialog box (see Figure 3.18). Note that the *Disable radius compensation* box is selected (circled in Figure 3.18), indicating that the radius compensation is currently disabled. Click the box again to enable radius compensation. Click *OK* to accept the change.

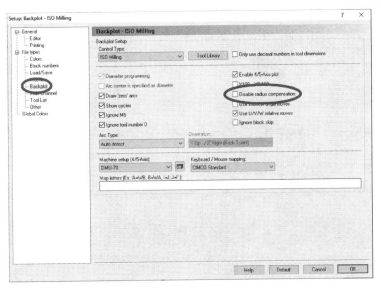

Figure 3.18 The *Setup: Backplot* dialog box

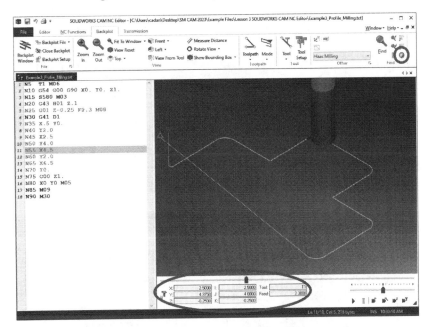

Figure 3.19 Toolpath with tool radius compensation

The toolpath with radius compensation appears in the backplot window (see Figure 3.19). Note that the toolpath is basically offset outward from the program path by the amount of the cutter radius, 0.375in. In addition, circular motions around points P2, P4, P5, and P7 are added to the toolpath. Click an NC block, for example line 11 (N55 X4.5), to move the cutter around point P4 (X: 2.5, Y: 4.0) following a circular path. The locations of the cutter are shown in the *Information Bar* below the backplot window, indicating X: 2.50, Y: 4.375, Z: –0.25, I: 2.50, J: 4.00, and K: –0.25. The toolpath shown in the backplot window now includes the radius compensation.

Machining Time

For this simple profile milling, we calculate the machining time manually.

The machining time can be calculated by dividing the distance that the cutter travels with the feedrate and add up the time of cutter retract.

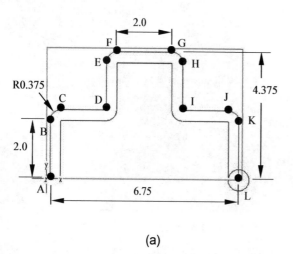

Points	Distance (inch)
AB	2
BC	0.375 (0.5π)
CD	2–0.375
DE	2–0.375
EF	0.375 (0.5π)
FG	2
GH	0.375 (0.5π)
HI	2–0.375
IJ	2–0.375
JK	0.375 (0.5π)
KL	2
Retract	1
Total	17.86

(a) (b)

Figure 3.20 Toolpath length, (a) sketch of the toolpath, and (b) the cutter travel distance

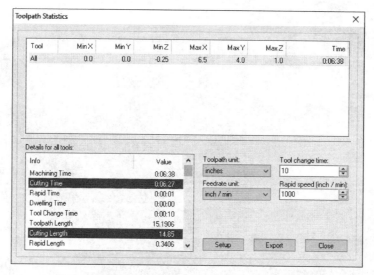

Figure 3.21 The *Toolpath Statistics* dialog box

The total distance d that the cutter travels, from points A to L shown in Figure 3.20(a), is 17.86 in., as shown in Figure 3.20(b). The feed rate f is 2.3 inch/minute; hence the cutting time is, without considering tool retract and non-cutting motion,

Machining time $= d/f = 17.86/2.3 = 7.77$ minutes

Click the *Toolpath Statistics* button ⚙ of the *Find* group (circled in Figure 3.19) to bring up the *Toolpath Statistics* dialog box like that of Figure 3.21. Note that the *Cutting Time* listed is *0:06:38*, which is close but less than our calculation. The *Cutting Length* is *14.85* in., which is closer to our calculation.

Click the *Editor* tab and select *Close* to close the NC program. We have finished the third example.

3.8   Example 4: Turn Finish Operation

In this example, we verify a turning NC program. This NC program is part of turning operations, see Figure 3.22(c), that cut a stepped bar [Figure 3.22(a)] from a round stock shown in Figure 3.22(b).

The major portion of the toolpath of the turning operations is machining the OD feature, consisting of *Turn Rough* and *Turn Finish*, as shown in Figure 3.23(a) and Figure 3.23(b), respectively. The turn rough toolpath moves the cutter towards the outer profile of the part in separate passes. The turn finish toolpath moves the cutter along the part profile in one pass, similar to the contour mill operation in milling. The cut off toolpath simply moves the cutter along the negative X direction to separate the part from the stock, as shown in Figure 3.23(c).

In this example, we only review the NC program of the *Turn Finish* operation shown in Figure 3.23(b). A complete turning operation can be found in Lesson 10 of this book.

Note that the cutter chosen for the turn finish operation is an insert of radius 0.02in. with an $80°$ angle, and the diameter of the inscribed circle is 0.5in. The cutter is following the silhouette profile boundary of the design model from points A to F shown in Figure 3.24, with an offset of the nose radius of the cutter; i.e., 0.02in. The X- and Z-coordinates of characteristic points A to F are listed in Table 3.1.

The NC program that performs this turn finish operation is shown in Figure 3.25. Note that this NC program shown in Figure 3.25 was created by using the post-processor of SOLIDWORKS CAM in Lesson 10, in which the X locations of the cutter are output as diametral instead of radial.

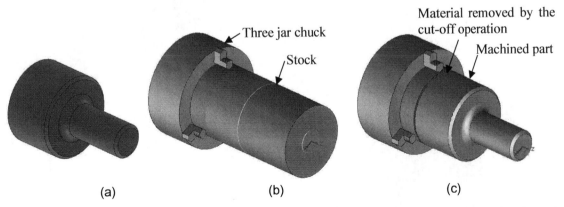

Figure 3.22 The stepped bar example, (a) the design model, (b) bar stock clamped to a chuck, and (c) the material removal simulation

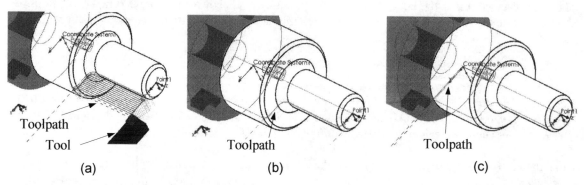

Figure 3.23 The toolpath of the turning operations, (a) *Turn Rough*, (b) *Turn Finish*, and (c) *Cut Off*

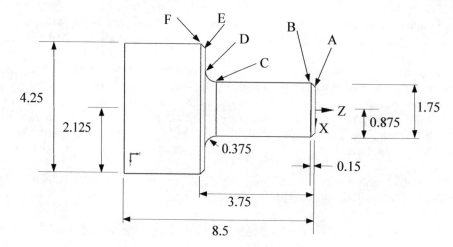

Figure 3.24 The six characteristic points (A to F) of the part boundary

Table 3.1 The XZ coordinates of the six characteristic points A to F

Point	X and Z Coordinates of Part Boundary	X and Z Coordinates of Toolpath	Offsets in X and Z Coordinates
A	X = 0.725 Z = 0	X = 0.7391 Z = 0.0141	ΔX = 0.0141 ΔZ = 0.0141
B	X = 0.875 Z = –0.15	X = 0.895 Z = –0.15	ΔX = 0.02 ΔZ = 0
C	X = 0.875 Z = –3.375	X = 0.895 Z = –3.375	ΔX = 0.02 ΔZ = 0
D	X = 1.25 Z = –3.75	X = 1.25 Z = –3.73	ΔX = 0 ΔZ = 0.02
E	X = 1.975 Z = –3.75	X = 1.975 Z = –3.73	ΔX = 0 ΔZ = 0.02
F	X = 2.125 Z = –3.9	X = 2.1391 Z = –3.8859	ΔX = 0.0141 ΔZ = 0.0141

We follow the same steps as previous examples to open *Example4_Turn_Finish.txt*. We select *ISO Turning*, select the *Backplot* tab, and click the *Backplot Window* button to show the backplot window like that of Figure 3.26. The backplot window of NC Editor for turning operations is not as desirable. First, the cutter is not quite visible and is not shown with correct cutter geometry. There is no tool holder, and no part boundary. No stock is visible either. Close the NC program to complete the example.

NC Commands	Explanations
O0001	Program number
N1 (DNMG 431 55DEG SQR HOLDER)	
N2 T01	Load Tool#1
N4 G00 G96 S374 M03	Turn on spindle
N5 (Turn Finish1)	
N6 G54 G00 Z.1241 M08	Set work coordinate system, move cutter to Z.1241
N7 X1.4783	Move cutter to X1.4783
N8 G01 X1.4783 Z.0241 F.0099	Turn on feedrate, move cutter to X1.4783 Z.0241
N9 Z.0141	Move cutter to Z.0141 (point A)
N10 X1.7783 Z-.1359	Move cutter to X1.7783 Z-.1359 (in front of point B)
N11 G03 X1.79 Z-.15 R.02	Move cutter to X1.79 Z-.15 R.02, circular motion (around point B)
N12 G01 Z-3.375	Move cutter to Z-3.375 (point C)
N13 G02 X2.5 Z-3.73 R.355	Move cutter to X2.5 Z-3.73 (point D)
N14 G01 X3.95	Move cutter to X3.95 (point E)
N15 G03 X3.9783 Z-3.7359 R.02	Move cutter to X3.9783 Z-3.7359 (around point E)
N16 G01 X4.2783 Z-3.8859	Move cutter to X4.2783 Z-3.8859 (point F)
N17 X4.2983	Move cutter to X4.2983 (around point F)
N18 G00 X4.4983	Move cutter to X4.4983
N19 G40 X6. Z1. M09	Move cutter to X6. Z1.
N20 M30	

Figure 3.25 Example 4: Finish turning, NC program with explanations

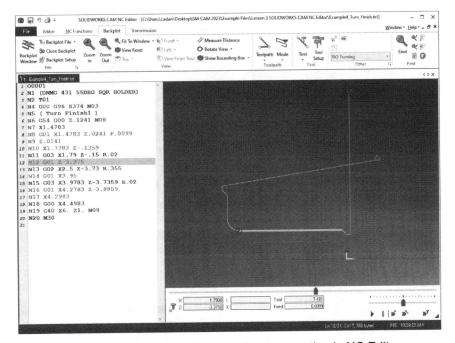

Figure 3.26 Toolpath of the turn finish operation in NC Editor

There are a few G-code viewers or editors available for support of NC program editing and verification, such as CNCSimulator Pro. They are available for free download for a prescribed trial period.

3.9   G-code Viewers and Editors

In addition to SOLIDWORKS CAM NC Editor, there are plenty of G-code viewers and editors available. Some are web-based on-line viewers and editors with which you may simply copy and paste NC blocks and display toolpath without downloading and installing software. This includes NC Viewer and G-Code Q'n'dirty toolpath simulator. Other viewers/editors require software download and install. Some require user's registration to acquire a login and password to activate the software. Most software offers a short-term trial period for free. A list of G-code viewers and editors are listed in Table 3.2 for your reference. In this section, we briefly discuss two such tools, NC Viewer, and CNCSimulator Pro. Please note that these web sites were accessed at the time the manuscripts were prepared. You may expect changes while accessing these sites.

Table 3.2 G-code viewers and editors

Software Name	Web Link*	Description	Remarks
CIMCO Edit v5	www.cimco.com www.cimco.com/download/registration/?p=edit&v=8.04.04	CIMCO Edit is one of the most popular CNC program editors. CIMCO Edit includes CNC code specific options such as line numbering / renumbering, character handling and XYZ range finder. It also features math functions including basic math, rotate, mirror, tool compensation, and translate.	Register to download. Demo version, can be used for 30 days
CNCSimulator Pro	cncsimulator.com	CNCSimulator Pro is a contemporary and advanced yet easy to use full 3D CNC simulation system with a virtual CNC controller and various machines as well as the integrated CAM system SimCam.	Register to download. Free for 30 days
NCPlot v2	ncplot.com/ncplotv2/download.htm	An editor and backplot window for 4 axis mill and 2 axis lathe G-code programs. This software combines editing, formatting and translation tools that are useful for CNC programmers with a backplot window for instant G-code verification	Register to download. Free for 15 days
NC Viewer v1.1.3	ncviewer.com	A G-code Viewer for Desktop and Mobile.	On-line, no software installation is needed
G-Code Q'n'dirty toolpath simulator	nraynaud.github.io/webgcode	A very simple G-code viewer on-line	On-line, no software installation is needed

* Web links accessed and verified before April 19, 2023.

NC Viewer

NC Viewer is a web-based on-line G-code viewer. You may go to the web link: ncviewer.com to start using the viewer without file downloading or software installing. On the web site, you may copy and paste the NC blocks you want to review on the left pane of the viewer (see Figure 3.27), and click the *PLOT* button (circled in Figure 3.27) to bring up the toolpath on the right pane. You may click the *Play* button ▶ to simulate or the *Step Forward* button ▶ to step through the toolpath.

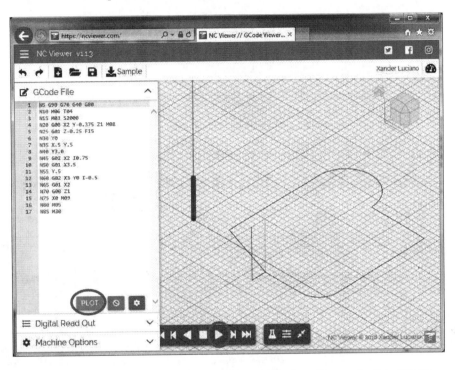

Figure 3.27 Toolpath of the trajectory milling operation (Example 1) in NC Viewer

CNC Simulator Pro

CNCSimulator Pro is a 3D CNC simulation system with a virtual CNC controller and machines as well as the integrated CAM system SimCam. You may use the software to define cutters, create workpiece, and set work coordinate system to simulate the toolpath of the NC program in both 2D and 3D. In 3D simulation, you will find various "virtual machines" that you can use to create realistic simulations with full 3D solids and many features of real CNC Machines. The software also supports zero point registry and tool libraries to radius compensations, sub programs and customized workpieces.

You may start CNCSimulator Pro, and open an NC program, in this case, the trajectory milling of Example 1. The NC program will appear in the right pane of the window similar to that of Figure 3.28 (with no toolpath on the left window yet).

Click the *Open the Inventory Browser* button 🖼 below the graphics area (circled in Figure 3.28) to bring up the *Inventory Browser* dialog box Figure 3.29(a). Under the *Tools* tab, click *Add* to add a cutter similar to that of Figure 3.29(b).

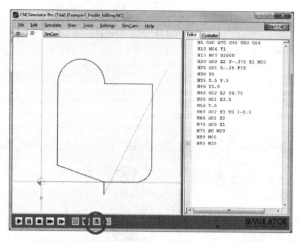

Figure 3.28 NC program of the trajectory milling operation (Example 1) listed in CNCSimulator Pro

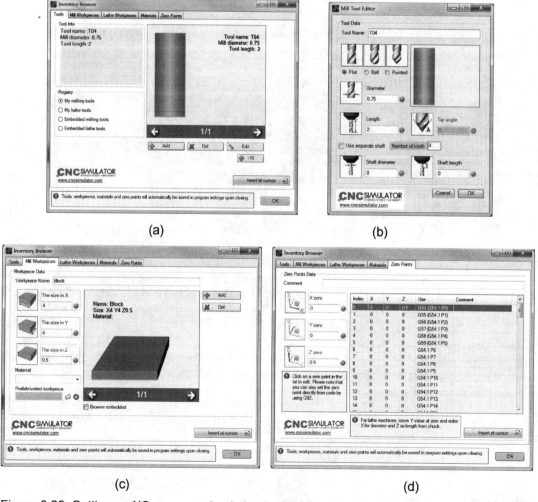

(a)   (b)

(c)   (d)

Figure 3.29 Setting up NC program simulation in CNCSimulator Pro, (a) the *Inventory Browser* dialog box, (b) the *Mill Tool Editor* dialog box, (c) The *Mill Workpieces* tab, and (d) the *Zero Points* tab

Figure 3.30 NC program simulation in CNCSimulator Pro in a more realistic setup under *3D* tab with
a virtual machine

Click the *Mill Workpieces* tab to enter dimension for the workpiece—4×4×0.5; see Figure 3.29(c), and
click the *Zero Points* tab to select the first coordinate system (first row) and enter 0, 0.5, and 0.5 for X, Y,
and Z respectively as the program zero; see Figure 3.29(d). Click *OK* to accept the setup in inventory.

Choose the *2D* tab above the graphics window, and click the *Play* button at the bottom to play the
toolpath. The toolpath appears in the graphics window similar to that of Figure 3.28. You may click *3D*
tab for a more realistic setup that includes a virtual machine like that of Figure 3.30.

3.10 Exercises

Problem 3.1. Verify the NC program you composed for **Problem 2.1** using:

(a)   SOLIDWORKS CAM NC Editor, and
(b)   CNCSimulator Pro.

Problem 3.2. Verify the NC program you composed for **Problem 2.2** using:

(a)   SOLIDWORKS CAM NC Editor, and
(b)   NC Viewer.

Problem 3.3. Verify the NC program of the turn finish operation of Example 4 using CNCSimulator Pro.

[Notes]

Lesson 4:   A Quick Run-Through

4.1   Overview of the Lesson

This lesson provides you with a quick start in using SOLIDWORKS CAM. You will learn a complete process in using SOLIDWORKS CAM to create a machining simulation from the beginning all the way to the post process that generates G-code for CNC machining. We will use a 3-axis mill to machine a simple plate by carrying out a contour milling operation (often called profile milling in other CAM software) that cuts the boundary profile of the part, as shown in Figure 4.1. You may want to open the model file, *Simple Plate with Toolpath.SLDPRT*, to preview the machining operation generated for this lesson. This lesson is intentionally made simple for new SOLIDWORKS CAM users. We stay with default options and machining parameters for most of the selections.

Figure 4.1 The material removal simulation of the simple plate example

We follow the general steps shown in Figure 1.1 of Lesson 1 to create machining simulation and generate toolpath using SOLIDWORKS CAM. In addition, we will go over a post process to generate G-code for this contour milling operation. We verify the accuracy of the G-code generated at the end.

After completing this lesson, you should be able to carry out machining simulation and toolpath generation for similar applications that involve contour milling operations following the same process. This lesson should also prepare you for the remaining lessons of the book.

4.2   The Simple Plate Example

The L-shape plate has a bounding box of size 4in.×4in.×0.5in., as shown in Figure 4.2, with a fillet of 0.375in. in radius. The unit system chosen is IPS (inch, pound, second). There is one solid feature created in the design model, *Boss-Extrude1*, listed in the FeatureManager design tree 🍀 (or simply model tree, feature tree, or SOLIDWORKS browser) shown in Figure 4.3. In addition, a coordinate system (*Coordinate System1*) is defined at the front left corner of the top face of the part. When you open the solid model *Simple Plate.SLDPRT*, you should see the solid feature and the coordinate system listed in the feature tree like that of Figure 4.3.

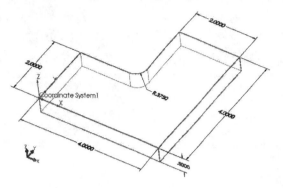

Figure 4.2 Dimensions of the simple plate model

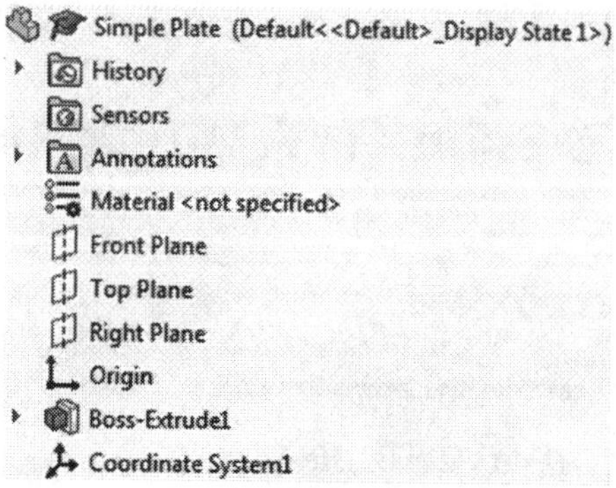

Figure 4.3 Entities listed in the feature tree

A stock of retangular block with a size 4.5in.×4.5in.×0.5in., made of low carbon alloy steel (1005), as shown in Figure 4.4, is chosen for the machining operation. Note that a part setup origin is defined at the front left vertex of the top face of the stock, which locates the G-code program zero.

We will create one contour milling operation, which cuts along the boundary profile of the part using a 1-in. flat-end mill. We will create the machinable feature interactively and follow the recommendations of the technology database (TechDB™) for choosing machining operation and parameters, such as feedrate and spindle speed. We will make a few changes to obtain a toolpath of the contour mill operation like that of Figure 4.5, in which the cutter moves along the boundary profile of the part in two passes. This is due to the fact that the machining depth (or depth of cut), 0.3in., chosen is less than the thickness of the stock (0.5in.).

4.3 Using SOLIDWORKS CAM

Open SOLIDWORKS Part

Open the part file (filename: *Simple Plate.SLDPRT*) downloaded from the publisher's website. This solid model, as shown in Figure 4.2, consists of one solid feature and a coordinate system among other entities. As soon as you open the model, you may want to check the unit system chosen.

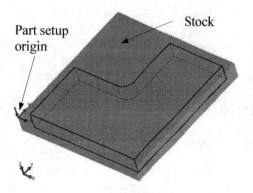

Figure 4.4 Stock with a part setup origin at its top left vertex

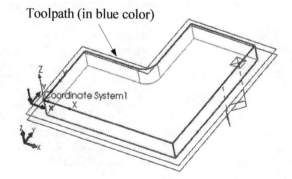

Figure 4.5 The toolpath of the contour milling operation

You may select from the pull-down menu *Tools > Options*. In the *Document Properties* dialog box (Figure 4.6), select the *Documents Properties* tab and select *Units*. Select *IPS (inch, pound, second)*. We will stay with this unit system for this lesson. Also, it is a good practice to increase the decimals from the default 2 to 4 digits since some machining parameters defined are down to a thousandth of an inch in SOLIDWORKS CAM. To increase the decimals to 4 digits, you may click the dropdown button to the right of the cell in the *Length* row under the *Decimals* column (circled in Figure 4.6). Pull down the selection and choose *.1234* for 4 digits. Click *OK* to accept the change.

Enter SOLIDWORKS CAM

At the top of the feature tree, you should see two important tabs, SOLIDWORKS CAM feature tree and SOLIDWORKS CAM operation tree, as circled in Figure 4.7. You may click these two tabs to review SOLIDWORKS CAM machinable features and machining operations, respectively. You should also see SOLIDWORKS CAM command buttons like those of Table 1.1 above the graphics area (by clicking the SOLIDWORKS CAM tab).

We will manually create a machinable feature (instead of clicking the *Extract Machinable Features* button), and use the four buttons next to it, *Generate Operation Plan* , *Generate Toolpath* , *Simulate Toolpath* , and *Step Through Toolpath* , to create and review the toolpath.

Before going through these steps, we will first select an NC machine and define a stock to be employed for machining the part.

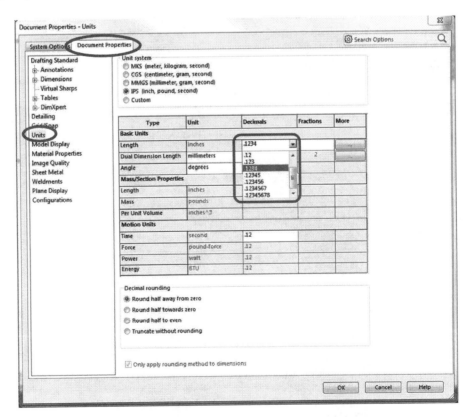

Figure 4.6 The *Document Properties* dialog box

Select NC Machine

Click the SOLIDWORKS CAM feature tree tab 🖼. A default mill *Mill-inch*, which is a 3-axis mill of inch system, is listed in the feature manager window (see Figure 4.8). This is the NC machine we want to use. We right click *Mill-inch* and choose *Edit Definition*.

SOLIDWORKS CAM
feature tree tab

SOLIDWORKS CAM
operation tree tab

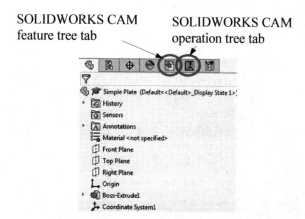

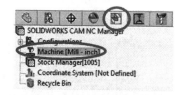

Figure 4.8 The NC machine
Mill-inch chosen by default

Figure 4.7 The SOLIDWORKS CAM feature tree
and SOLIDWORKS CAM operation tree tabs

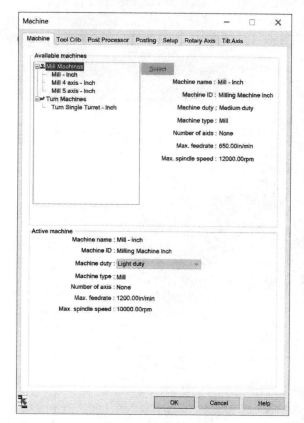

Figure 4.9 The *Machine* tab of the *Machine*
dialog box

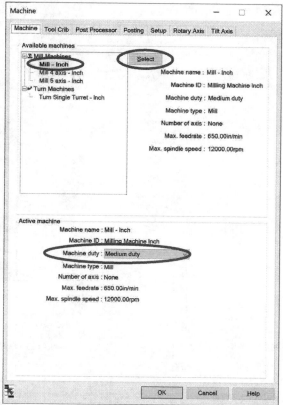

Figure 4.10 The *Mill-inch* machine selected and
highlighted in bold face

In the *Machine* dialog box (see Figure 4.9), available machines are listed in the box below *Available machines* under *Machine* tab. *Mill-inch* (along with other two mills) is listed under *Mill Machines*. A few basic machine parameters of the mill *Mill-inch* (for example *Max. feedrate*) are also listed. This is because by default *Mill-inch* (light duty) is active (see information under *Active machine*).

Choose *Mill-inch* in the box under *Available machines*, and click *Select. Mill-inch* is now highlighted in bold face, as shown in Figure 4.10. Select *Medium Duty* for *Machine duty* (circled in Figure 4.10).

Next, we will choose *Tool Crib*, *Post Processor*, and *Setup* tabs to review or modify the machine definition.

Choose the *Tool Crib* tab and select *Tool Crib 2* under *Available tool cribs* (see Figure 4.11), and then click *Select*. Tools available in *Crib 2* are now listed. We will use the tools available in *Crib 2* for this lesson.

Choose the *Post Processor* tab; a post processor called *M3AXIS-TUTORIAL* is selected (see Figure 4.12). This is a generic post processor of 3-axis mill that comes with SOLIDWORKS CAM.

There are other post processors that come with SOLIDWORKS CAM, which are located in *C:\ProgramData\SOLIDWORKS\SOLIDWORKS CAM 2023\Posts*. Note that in practice you will have to identify a suitable post processor that produces G-code compatible with the CNC machines at the shop floor. We will stay with *M3AXIS-TUTORIAL* for this lesson.

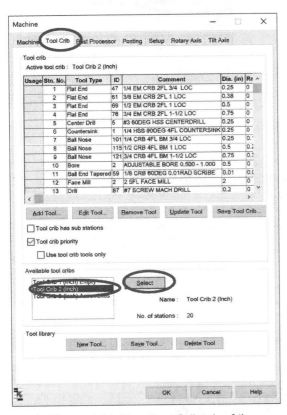

Figure 4.11 The *Tool Crib* tab of the *Machine* dialog box

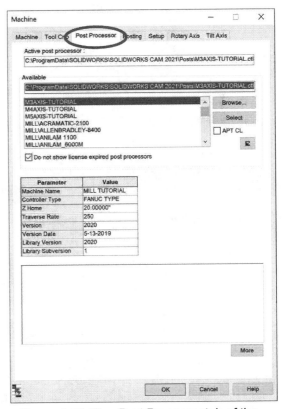

Figure 4.12 The *Post Processor* tab of the *Machine* dialog box

Choose the *Setup* tab, and click the *Define* button; see Figure 4.13(a). In the *Fixture Coordinate System* dialog box, choose *SOLIDWORKS Coordinate System*, circled in Figure 4.13(b), and click *Coordinate System1*. The selected coordinate system is now listed under *Selected Coordinate System*; see Figure 4.13(b).

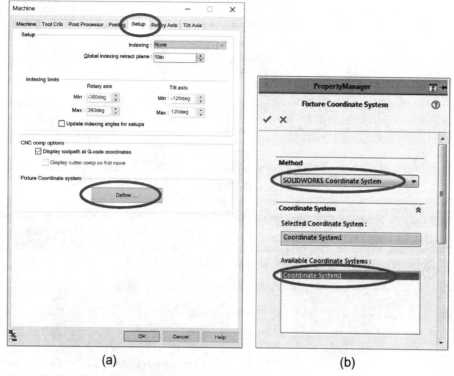

| (a) | (b) |

Figure 4.13 The *Setup* tab of the *Machine* dialog box, (a) choosing *Edit* tab, and clicking *Define*, and (b) choosing *Coordinate System1* for *Fixture Coordinate System*

Note that in SOLIDWORKS CAM, the fixture coordinate system defines the "home point" or main zero position on the machine. For some machining conditions, defining the fixture coordinate system is optional. However, to be safe, it is recommended that you always define an FCS.

Click *OK* to accept the selections and close the dialog box.

Create Stock

From SOLIDWORKS CAM feature tree 🗒, right click *Stock Manager* and choose *Edit Definition* (see Figure 4.14). The *Stock Manager* dialog box appears (Figure 4.15), in which a default stock size appears at the bottom of the dialog box, which is the size of the bounding box of the part. The default stock material is *1005* (if not, select *1005* for stock material). We will increase the length and width of the stock by 0.25in. on both sides (enter 0.25 for *X+*, *X-*, *Y+*, and *Y-*, as shown in Figure 4.15). The stock size becomes X:4.5, Y:4.5, Z:0.5, as circled in the *Stock Manager* dialog box (Figure 4.15).

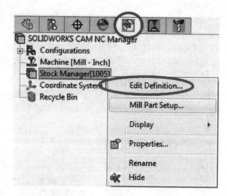

Figure 4.14 Right clicking *Stock Manager* and choosing *Mill Part Setup* from the feature tree

Accept the revised stock by clicking the checkmark ✔ at the top left corner. The rectangular stock should appear in the graphics area similar to that of Figure 4.4.

Mill Part Setup and Machinable Feature

Since a machinable feature of the contour milling is not extractable by SOLIDWORKS CAM automatically, we will create one manually. We will first create a mill part setup and then insert a 2.5 axis feature.

From the SOLIDWORKS CAM feature tree 📷, right click *Stock Manager* and choose *Mill Part Setup*. The *Mill Setup* dialog box appears (Figure 4.16) with the *Front Plane* selected under *Entity*. In the graphics area (similar to that of Figure 4.17), a tool axis symbol ⊕ with arrow pointing upward appears. This symbol indicates that the tool axis (or the feed direction) must be reversed (pointing downward).

Click the *Reverse Selected Entity* button 🔃 under *Entity* (circled in Figure 4.16) to reverse the direction. Make sure that the arrow points in a downward direction like ⬇. Click the checkmark ✔ to accept the definition. A *Mill Part Setup1* is now listed in the SOLIDWORKS CAM feature tree, as shown in Figure 4.18.

Next we define a machinable feature. From SOLIDWORKS CAM feature tree 📷, right click *Mill Part Setup1* just created and choose *2.5 Axis Feature*.

In the *2.5 Axis Feature* dialog box (Figure 4.19), choose *Boss* for *Type*, and select *Sketch1* under *Available Sketches*, then click the *Next* button 🔵 (circled in Figure 4.19) to define end condition.

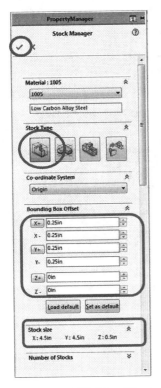

Figure 4.15 The *Stock Manager* dialog box

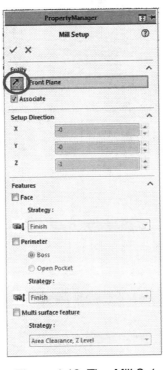

Figure 4.16 The *Mill Setup* dialog box

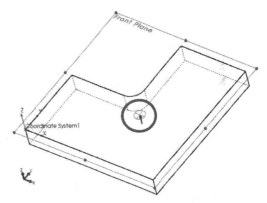

Figure 4.17 The *Front Plane* selected

Figure 4.18 *Mill Part Setup1* listed in the SOLIDWORKS CAM feature tree

Note that *Sketch1* defines the outer profile of the plate. Choose *Finish* (default) for *Strategy* (circled in Figure 4.20), and enter *0.5* for the depth dimension under *End condition – Direction 1*, then click the checkmark ✔ to accept the definition. An *Irregular Boss1* node is now listed in the SOLIDWORKS CAM feature tree 🖼 in magenta color (Figure 4.21), indicating that the machining feature is unfinished.

Generate Operation Plan and Toolpath

Right click *Irregular Boss1* and choose *Generate Operation Plan* (or click the *Generate Operation Plan* button ⬚ᴳᵉⁿᵉʳᵃᵗᵉ above the graphics area). A new node, *Contour Mill1*, is listed in SOLIDWORKS CAM operation tree 🔳 (see Figure 4.22) in magenta color. Note that you are now shifted automatically to SOLIDWORKS CAM operation tree 🔳.

The NC operation, *Contour Mill1*, is assigned by SOLIDWORKS CAM by selecting a suitable operation in the TechDB™, in which a 0.5in. diameter flat-end cutter is chosen. Click the node under operation tree tab, *Contour Mill1*, to display a summary of the NC operation at the top left corner in the graphics window, including tool material, XY and Z feedrates, and machining depth (0.5in.), as shown in Figure 4.23. We will stay with default setups and parameter values for the time being. Later in this lesson, we learn to choose a different cutter and modify some of the parameters for a toolpath we intend to generate.

Right click *Contour Mill1* and choose *Generate Toolpath* (or click the *Generate Toolpath* button 🔧ᴳᵉⁿᵉʳᵃᵗᵉ Toolpath above the graphics area).

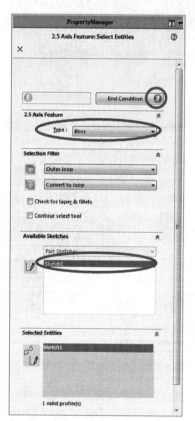

Figure 4.19 The *2.5 Axis Feature* dialog box

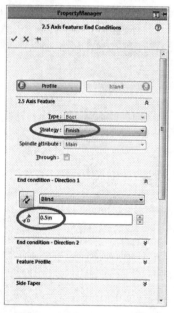

Figure 4.20 Defining end conditions

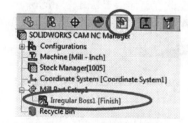

Figure 4.21 Machinable feature added to the feature tree

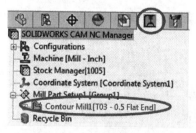

Figure 4.22 Machining operation added to the operation tree

A contour mill toolpath will be generated like that shown in Figure 4.24. Note that the flat-end cutter cuts along the part profile in two passes; the depth of cut was chosen as 0.25in. by SOLIDWORKS CAM.

Define Part Setup Origin

The toolpath generated, as shown in Figure 4.24, assumes a part setup origin (see the symbol circled in Figure 4.24) located at the front left corner of the design part, coinciding with the fixture coordinate system, *Coordinate System1*. You may see the part setup origin symbol more clearly by right clicking the coordinate system node, *Coordinate System1*, from the FeatureManager design tree and choosing *Hide* to hide it.

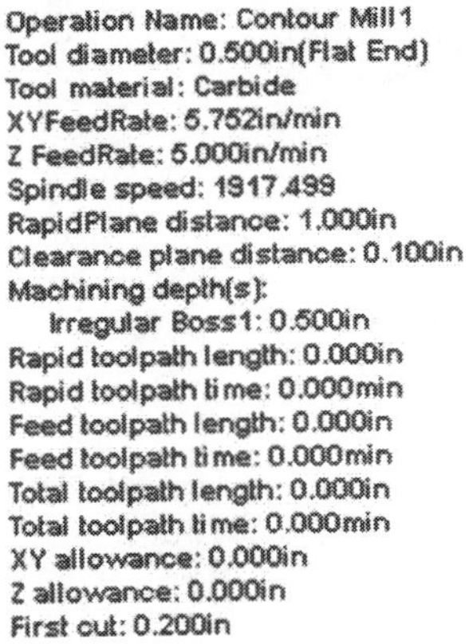

Operation Name: Contour Mill1
Tool diameter: 0.500in(Flat End)
Tool material: Carbide
XYFeedRate: 5.752in/min
Z FeedRate: 5.000in/min
Spindle speed: 1917.499
RapidPlane distance: 1.000in
Clearance plane distance: 0.100in
Machining depth(s):
Irregular Boss1: 0.500in
Rapid toolpath length: 0.000in
Rapid toolpath time: 0.000min
Feed toolpath length: 0.000in
Feed toolpath time: 0.000min
Total toolpath length: 0.000in
Total toolpath time: 0.000min
XY allowance: 0.000in
Z allowance: 0.000in
First cut: 0.200in

Figure 4.23 A suitable operation found in TechDB™

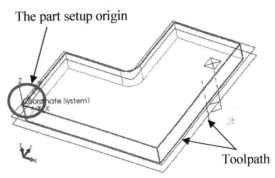

The part setup origin
Toolpath

Figure 4.24 The toolpath generated

As mentioned before, the part setup origin defines the program zero for the G-code to be generated after the toolpath is finalized. A part setup origin defined at the corner of the design part is less desirable. In practice, it is more convenient to set the origin at a corner of the stock, for example, its top left corner, which is accessible and easier to set up on the mill.

We right click *Mill Part Setup* under the SOLIDWORKS CAM operation tree tab and choose *Edit Definition*. In the *Part Setup Parameters* dialog box (Figure 4.25), select the *Origin* tab, choose *Stock vertex* and pick the vertex at the front corner of the top face of the sample stock (circled in Figure 4.25). The part setup origin in the graphics area is now moved to the front left corner of the top face of the stock like that of Figure 4.4. Click *OK* to accept the change.

Part Setup Parameters
Origin
Offset
Indexing
Advanced
Statistics
NC Planes
Posting
Origin
User
X: -0.25in
Y: -0.25in
Z: 0in
Entity
Sketch
Sketch1
Fixture coordinate system
Top center (Part model)
Mid center (Part model)
Bottom center (Part model)
Stock vertex
OK
Cancel
Help

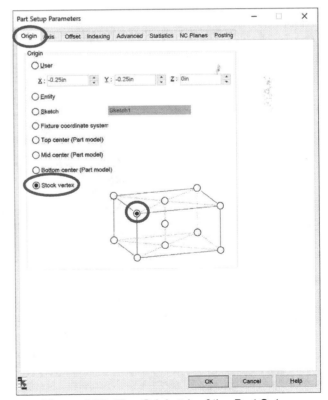

Figure 4.25 The *Origin* tab of the *Part Setup Parameters* dialog box

Click the *Yes* button in the SOLIDWORKS CAM warning box: *The origin or machining direction or advanced parameters has changed, toolpaths need to be recalculated. Regenerate toolpaths now?* The toolpath will be regenerated referring to where the setup origin is relocated.

Modify the Toolpath

Next we replace the tool with a 0.75in. flat-end mill, and increase the machining depth to 0.3in.

Under the SOLIDWORKS CAM operation tree tab ![icon], right click *Contour Mill1* and choose *Edit Definition*. In the *Operation Parameters* dialog box, choose *Tool* tab; the 0.5in. flat-end cutter is shown (Figure 4.26). Click the *Tool Crib* tab, select the 4th tool (Type: *FLAT END*, ID: *76*, Comment: *3/4 EM CRB 2FL 1-1/2LOC*, implying 3/4in. End Mill of Solid Carbide with 2 Flutes and 1.5in. Length of Cut), click the *Select* button to select the tool (see Figure 4.27).

Click *Yes* to the question in the warning box: *Do you want to replace the corresponding holder also?*

Choose the *Contour* tab of the *Operation Parameters* dialog box. In the *Depth* parameters group, click the percentage button ![icon] to deselect it for both *First cut amt.* and *Max cut amt.* Enter *0.3in.* for both *First cut amt.* and *Max cut amt.*, as shown in Figure 4.28, since a larger tool has been chosen.

Click the *F/S* tab to review machining parameters (Figure 4.29); for example, the *XY feedrate* is chosen as *4.437in./min.* by the techology database of SOLIDWORKS CAM. Click *OK* to accept the changes.

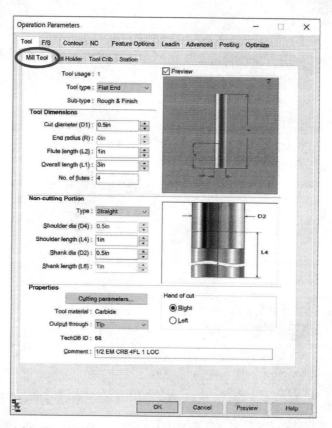

Figure 4.26 The *Mill Tool* tab of the *Operation Parameters* dialog box

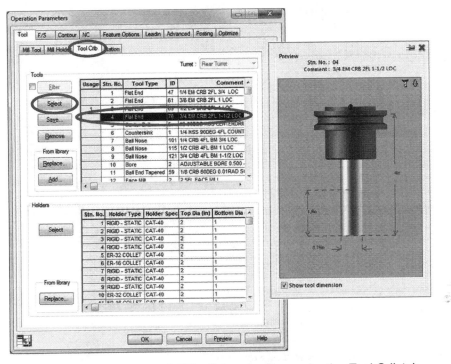

Figure 4.27 Selecting a 0.75in. flat-end cutter under the *Tool Crib* tab

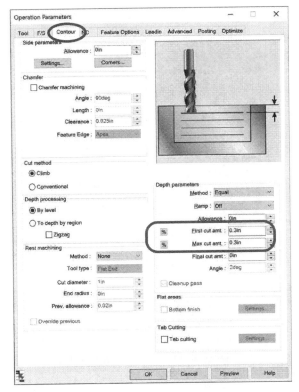

Figure 4.28 Defining depth parameters under the *Contour* tab

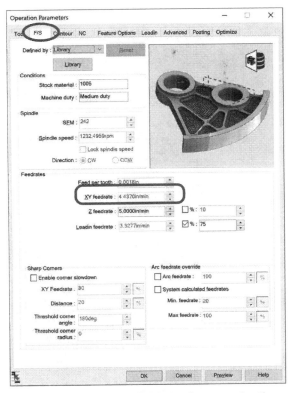

Figure 4.29 The *XY feedrate* shown under the *F/S* tab

Right click *Contour Mill1* node and choose *Generate Toolpath*. Profile milling toolpath will be generated like that shown in Figure 4.5 with two passes.

Simulate Toolpath

Right click *Contour Mill1* node and choose *Simulate Toolpath* (or click the *Simulate Toolpath* button above the graphics area). The *Toolpath Simulation* tool box appears (Figure 4.30).

Click the *Run* button circled in Figure 4.30 to simulate the toolpath via a material removal simulation. The machining simulation of the contour milling operation will appear in the graphics area, similar to that of Figure 4.1.

Step Through Toolpath

You may step through the toolpath to learn more about the individual machining steps. Right click *Contour Mill1* node and choose *Step Thru Toolpath* (or click the *Step Thru Toolpath* button above the graphics area). The *Step Through Toolpath* dialog box appears (Figure 4.31). Under *Information*, SOLIDWORKS CAM shows the tool movement from the current to the next steps (in X, Y, and Z coordinates), the feedrate and spindle speed, among others.

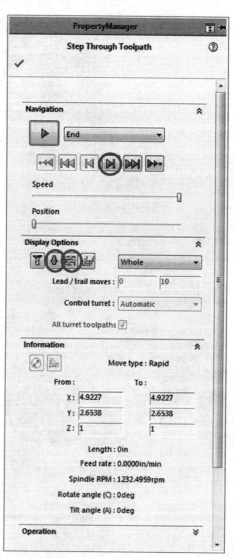

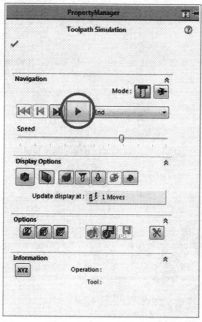

Figure 4.30 Playing the machining simulation by clicking the *Run* button

Figure 4.31 The *Step through Toolpath* dialog box

Click *Step* button ⏭ at the center of the tool box (circled in Figure 4.31) to step through the toolpath. You may want to turn on *Show toolpath points*, and *Tool Holder Shaded Display* (circled in Figure 4.31) to see the toolpath displayed on the part similar to that of Figure 4.32.

4.4   The Post Process and G-code

Now we have completed the contour milling operation. Next, we learn how to convert the toolpath to G-code and how to create a CL data file.

Right click *Contour Mill1* node and choose *Post Process* (or click the *Post Process* button 🔘 above the graphics area). In the *Post Output File* dialog box (Figure 4.33), choose a proper file folder, enter a file name (default name is *Simple Plate.txt*), use the default file type (*M3Axis-Tutorial*), and click *Save*. The *Post Process* dialog box appears (Figure 4.34).

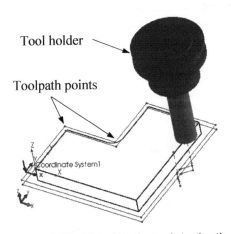

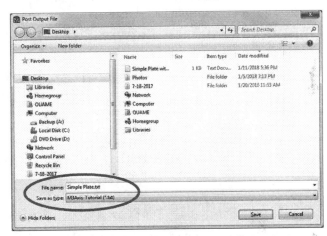

Figure 4.32 Stepping through toolpath

Figure 4.33 The *Post Output File* dialog box

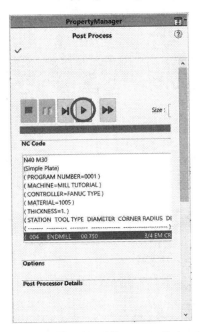

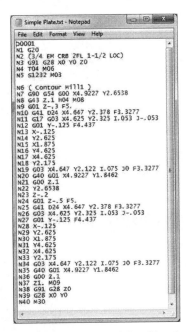

Figure 4.34 The *Post Process* dialog box

Figure 4.35 The contents of the *Simple Plate.txt* file

In the *Post Process* dialog box, click the *Play* button (circled in Figure 4.34) to create a G-code file (.txt file). Open the G-code file (filename: *Simple Plate.txt*) from the folder using *Word* or *Word Pad* (see the file contents shown in Figure 4.35 with command explanations provided in Column B of Table 4.1).

Note that the NC word G41 in blocks N10 and N25 turns on tool radius compensation from the left for Tool#24 (since D24 is included in the NC block), which moves the tool to the left of the program path (that is, the boundary profile of the design model) by the amount of the radius of Tool#24. Since the cutter locations of the G-code specify the actual locations of the tool already (such as Points A, B, C, etc. shown in Figure 4.39), tool radius compensation is not necessary. In fact, Tool#24 was never loaded. G41 D24 in Blocks N10 and N25 are not needed at all. This is an error created by the post processor, *M3AXIS-TUTORIAL*, which must be corrected.

After correcting the error (and perhaps a few more, depending on the controller of your NC machine), the .txt file can be uploaded to a CNC mill, for example a HAAS mill, to machine the part. In general, the G-code converted is ready to be loaded to a mill. However, a few minor changes may be needed (or errors found in this example to be corrected), depending on the post processor employed in outputting the G-code output.

Click the *Save CL File* button ▦ Save CL File above the graphics area. The *Save As* dialog box appears (Figure 4.36).

In the *Save As* dialog box (Figure 4.36), choose a proper file folder, enter a file name (default name is *Simple Plate.clt*), use the default type (*CL Files*), and click *Save*. Open the CL file to review its contents (Figure 4.37). The GOTO indicates how the cutter moves along the profile of the part. The X-, Y-, and Z-coordinates of the cutter location are identical to those of the G-code.

4.5   Reviewing Machining Time

From SOLIDWORKS CAM operation tree ▦ , right click *Contour Mill1* and choose *Edit Definition*.

In the *Operation Parameters* dialog box (Figure 4.38), choose the *Optimize* tab, and look for *Estimated machining time*. The total tool length of feed is 40.8in., and the feed time is 9.314 minutes.

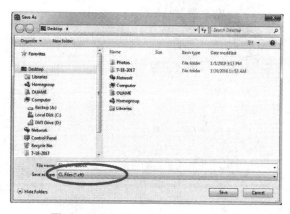

Figure 4.36 The *Save As* dialog box

```
UNITS/ INCHES
BOX/ 2.000000,2.000000,-0.250000,4.000000,4.000000,0.500000
CUTTER/ 0.750000,0.000000,0.375000,0.000000,0.000000,0.000000,4.000000,
TLAXIS/ 0.000000,0.000000,1.000000
OPFEATSTART/Contour Mill1-Irregular Boss1
ROTABL/ 0.000000,AAXIS,TABLE,CCLW
ROTABL/ -0.000000,BAXIS,TABLE,CCLW
ROTABL/ 0.000000,CAXIS,TABLE,CCLW
RAPID/
GOTO/ 4.922739,2.653805,1.000000
RAPID/
GOTO/ 4.922739,2.653805,0.100000
FEDRAT/ IPM,5.000000
GOTO/ 4.922739,2.653805,-0.300000
CUTCOM/ ON
CUTCOM/ LEFT
FEDRAT/ IPM,3.327739
GOTO/ 4.646967,2.378033,-0.300000
CIRCLE/ 4.700000,2.325000,-
0.300000,0.000000,0.000000,1.000000,0.075000,COUNTERCLOCKWISE
GOTO/ 4.625000,2.325000,-0.300000
FEDRAT/ IPM,4.436985
GOTO/ 4.625000,-0.125000,-0.300000
GOTO/ -0.125000,-0.125000,-0.300000
GOTO/ -0.125000,2.625000,-0.300000
GOTO/ 1.875000,2.625000,-0.300000
GOTO/ 1.875000,4.625000,-0.300000
GOTO/ 4.625000,4.625000,-0.300000
GOTO/ 4.625000,2.175000,-0.300000
FEDRAT/ IPM,3.327739
```

Figure 4.37 The partial contents of the
CL file (*Simple Plate.clt*)

Table 4.1 Contents of the .txt file (NC codes) with explanations

NC Commands (Column A)	Explanations (Column B)
O0001	Main program, program #0001
N1 G20	G20: Select inches
N2 (3/4 EM CRB 2FL 1-1/2 LOC)	Comments in parentheses
N3 G91 G28 X0 Y0 Z0	G91: incremental programming, G28: Return to reference point, move cutter to X0 Y0 Z0
N4 T04 M06	M06: Tool Change, Load tool #T04
N5 S1232 M03	M03: Spindle forward at speed 1232rpm
N6 (Contour Mill1)	Comments
N7 G90 G54 G00 X4.9227 Y2.6538	G90: Absolute programming, G54: Select Work Coordinate System 1, G00: Move cutter to start point (X4.9227 Y2.6538) rapidly, non-cutting
N8 G43 Z.1 H04 M08	G43: Tool length compensation with length registered for the cutter #04, M08: Coolant on
N9 G01 Z-.3 F5.	G01: Cutting, plunge the cutter 0.3in. into the stock at feedrate 5.0in./min
N10 **G41 D24** X4.647 Y2.378 F3.3277	G41: 2D cutter radius compensation from left *(this is an error, should be removed)*, D24: tool diameter registered at #24 *(this is an error since tool is T04, should be removed)*.
N11 G17 G03 X4.625 Y2.325 I.053 J-.053	G17: Cutting on the XY plane, G03: Cutting, circular motion moving the cutter to Point A at feedrate 2.5669in./min (see Figure 4.39 for the toolpath points)
N12 G01 Y-.125 F4.437	G01: Cutting, move cutter to Point B at feedrate 4.437in./min
N13 X-.125	Point C
N14 Y2.625	Point D
N15 X1.875	Point E
N16 Y4.625	Point F
N17 X4.625	Point G
N18 Y2.175	Point A': X4.625 Y2.175 (not shown in Figure 4.39)
N19 G03 X4.647 Y2.122 I.075 J0 F3.3277	G03: Cutting, circular motion moving the cutter to X4.647 Y2.122
N20 G40 G01 X4.9227 Y1.8462	G40: Cancel radius compensation (no need), G01: Cutting, linear motion moving cutter to X4.9227 Y1.8462
N21 G00 Z.1	Retract to 0.1in. in the Z-direction, rapid
N22 Y2.6538	Move to X4.9227 Y2.6538
N23 Z-.2	Move to -0.2 in. the Z-direction, rapid
N24 G01 Z-.5 F5.	G01: Cutting, plunge the cutter 0.5in. into the stock at feedrate 5.0 in./min for the second pass
N25 **G41 D24** X4.647 Y2.378 F3.3277	Same as block N10
N26 G03 X4.625 Y2.325 I.053 J-.053	Similar to Block N11 (Point A)
N27 G01 Y-.125 F4.437	Same as N12 (Point B)
N28 X-.125	Point C
N29 Y2.625	Point D
N30 X1.875	Point E
N31 Y4.625	Point F
N32 X4.625	Point G
N33 Y2.175	Point A'
N34 G03 X4.647 Y2.122 I.075 J0 F3.3277	Same as N19
N35 G40 G01 X4.9227 Y1.8462	Same as N20
N36 G00 Z.1	Retract to 0.1in. in the Z-direction, rapid
N37 Z1. M09	Retract to 1.0 in. in the Z-direction, rapid, M09: Coolant Off
N38 G91 G28 Z0	G91: Incremental programming, G28: Return to reference point, move cutter to Z0
N39 G28 X0 Y0	G28: Return to reference point, move cutter to X0Y0
N40 M30	Program end and rewind

The feed time can be manually calculated by dividing the distance that the cutter travels (see Figure 4.39) by the feedrate.

Since the total distance d that the cutter travels along the part boundary profile is approximately 38in. (19in. per pass and there are two passes), as tabulated in Table 4.2, and the feedrate f is *4.437* in./min. (see Figure 4.29), the feed time can be calculated as

Feed Time = d/f = 38/4.437 = 8.56 minutes

The feed time calculated is not exactly the same as estimated by SOLIDWORKS CAM since the feed in the vertical direction (for examples, N9 and N24 in Table 4.1) and before cutter reaching Point A (N10, N11, N19, N25, N26, N34) were not included. However, this value is close to that of SOLIDWORKS CAM; i.e., 9.372 minutes shown in Figure 4.38.

We have completed this tutorial lesson. You may save your model for future reference.

Table 4.2 Cutter travel distance in one pass with feedrate turned on

Point	Distance (in.)
AB	2.45
BC	4.75
CD	2.75
DE	2
EF	2
FG	2.75
GA	2.3
Total	19

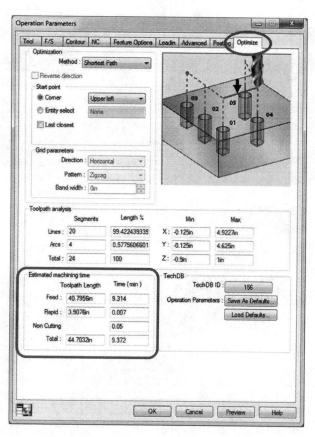

Figure 4.38 The *Optimize* tab of the *Operation Parameters* dialog box

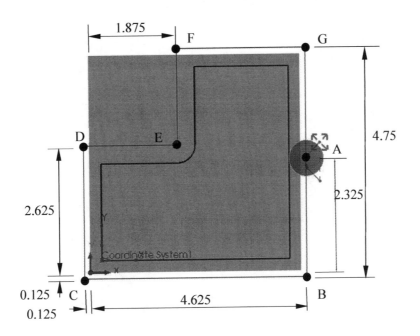

Figure 4.39 Dimensions of the cutter location points (labeled as A, B, C, and so on) and a shaded stock

4.6   Exercises

Problem 4.1.
In this lesson, we learned to create a 2.5 axis feature for the contour milling operation. In fact, there is another feature called *Part Perimeter Feature* that may also be suitable for this example in practice. You may create such a machinable feature by right clicking *Mill Part Setup1* and choosing *Part Perimeter Feature*. You are asked to create such a part perimeter feature and carry out a machining simulation following the process learned in this lesson. Review the toolpath generated. Compare the toolpath with that of this lesson. Identify major differences, pros, and cons between these two. Which toolpath is more realistic?

Problem 4.2. Create a contour milling operation using SOLIDWORKS CAM to machine a block for the design model shown in Figure 4.40 by creating a 2.5 axis feature. The part file can be found at the publisher's website. It is located in Lesson 4/Exercises folder. The stock is 7in.×4.5in.×0.5in. of steel 1005, as shown in Figure 4.40. The feedrate is assumed 5 in./minute, and the depth of cut is 0.5in. In this exercise, please report the following:

- The tool you chose;
- Feed time obtained from SOLIDWORKS CAM;
- Feed time obtained from your own calculations;
- Screen captures for toolpath and material removal simulation in SOLIDWORKS CAM.

Grade should be given based on the following:

- Quality of the toolpath, i.e., the machined part must be identical to the design model (a clean cut);
- Feed time, a least possible feed time is desired.

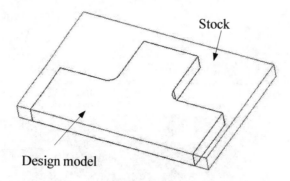

Stock

Design model

Figure 4.40 Design model and stock of Problem 4.2

Problem 4.3. Repeat Problem 4.2 for the design model shown in Figure 4.41. The stock is a rectangular block of 35in.×20in.×2in. of steel 1005. The feedrate is assumed 5 inch/minute, and the step depth is 0.5 inch.

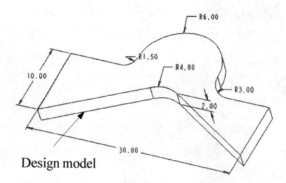

Design model

Figure 4.41 Design model of Problem 4.3

Lesson 5: Machining 2.5 Axis Features

5.1   Overview of the Lesson

In this lesson, we learn milling operations for machining 2.5 axis features. The major characteristic of a 2.5 axis feature is that the top and bottom of the feature are flat and are normal to the tool axis of the machining operations under the mill part setup. Such features include prismatic solid features and solid features with tapered walls. Typical 2.5 axis features include boss, pocket, open pocket, corner slot, slot, hole, face feature, open profile, curve or engrave feature. Some of these features are illustrated in Figure A.1 of Appendix A.

Features of this type are often represented as a profile sketch in CAD software that gives the height of the feature at individual characteristic points. 2.5 axis features are often preferred for machining, as it is easy to generate G-code for them in an efficient, often close to optimal fashion. In general, 2.5 axis features can be machined using a 3-axis mill.

Most 2.5 axis features, such as pockets, holes, slots, and bosses, can be automatically extracted as machinable features by using the automatic feature recognition (AFR) capability of SOLIDWORKS CAM. Others, such as face milling features, or contour (or profile) milling features (as seen in Lesson 4), can be created manually (or interactively) using Interactive Feature Recognition (IFR) capability. In this lesson, we will use both methods, AFR and IFR, to extract and select respective machinable features from a solid model, generate operation plans and toolpaths, simulate and step through toolpaths, and post process the toolpaths for G-code.

The part solid model (or design model) of this lesson shown in Figure 5.1(a) consists of six holes and a pocket. They are all 2.5 axis features and are extracted as machinable features automatically. In addition, we will create a face milling operation that removes a layer of material on top of the part. A material removal simulation is shown in Figure 5.1(b).

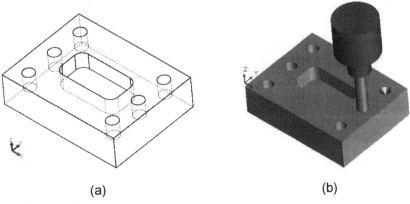

(a)   (b)

Figure 5.1 The 2.5 axis features example, (a) part solid model, and (b) material removal simulation

After completing this lesson, you should be able to generate machining simulations for similar parts that involve machining 2.5 axis features following the same procedures. In this exercise, we will use default options for most of the selections.

5.2   The 2.5 Axis Features Example

The size of the bounding box of the part (filename: *2 point 5 axis features.SLDPRT*) is 8in.×6in.×2in., as shown in Figure 5.2(a). The base block is created as a *Boss-Extrude1* solid feature. The size of the center pocket, *Cut-Extrude1*, is 4in.×2in.×1in. with fillets of 0.5in. in radius at the four corners; see Figure 5.2 (b). There are six blind holes, three on each side, of diameter 0.75in. and depth 1.0in.; see Figure 5.2 (c). The first hole was created as a cut extrude feature (*Cut-Extrude2*) and then duplicated for additional instances by using a linear pattern feature, as shown in Figure 5.2 (d).

In addition, a coordinate system (*Coordinate System1*) is created at the front left corner of the bottom face of the part. This coordinate system will be chosen as the fixture coordinate system, which defines the "home point" or main zero position on the machine. However, please note that the G-code generated refers to the part setup origin, as discussed in Lesson 4, which does not have to be the same as the fixture coordinate system. The unit system chosen is IPS (inch, pound, second). When you open the solid model: *2 point 5 axis features.SLDPRT*, you should see the four solid features and a coordinate system listed in the feature tree like that of Figure 5.3.

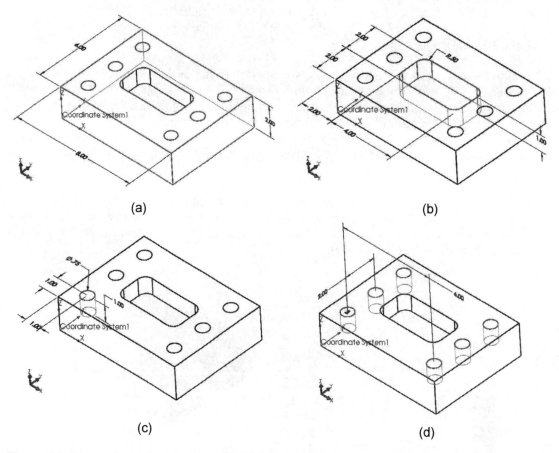

Figure 5.2 Dimensions of the solid features of the design model: (a) the base extrude, (b) the pocket, (3) the corner hole, and (d) the hole linear pattern

A stock of rectangular block with size 8in.×6in.×2.25in., made of low carbon alloy steel (1005), as shown in Figure 5.4, is chosen for the machining operations. A layer of 0.25in. above the top face of the part will be removed by creating a face milling operation. Note that a part setup origin is defined at the front left vertex at the top face of the stock, which locates the G-code program zero.

Figure 5.3 Feature tree of the example part

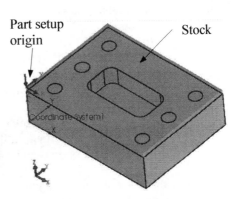

Figure 5.4 Stock with a part setup origin created at the front left vertex on top face

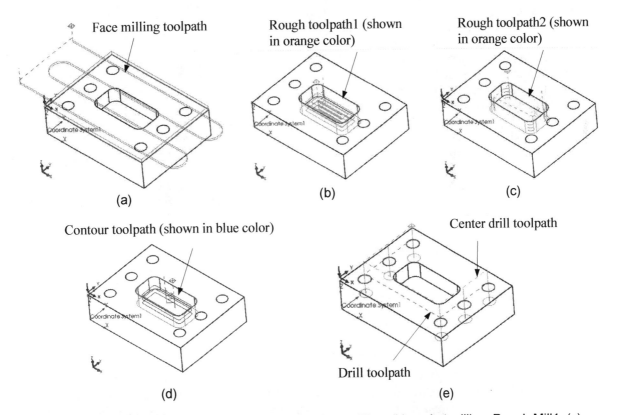

Figure 5.5 Toolpaths of the six NC operations: (a) face milling, (b) pocket milling: Rough Mill1, (c) pocket milling: Rough Mill2, (d) pocket milling: Contour Mill1, and (e) hole drilling: Center Drill1 and Drill1

There are six NC operations to be generated for this example. The first operation is face milling, which is a rough cut using a 2.5in. flat-end mill (T20 in the tool crib). The toolpath of the face milling is shown in Figure 5.5(a).

The second operation—see Figure 5.5(b)—is a rough mill operation that cuts the center pocket (*Cut-Extrude1* solid feature) using a 0.75in. flat-end mill (T04). The third—see Figure 5.5(c)—is another rough mill that cuts the corner fillets of the center pocket using a 0.5in. flat-end mill (T03). The fourth operation—see Figure 5.5(d)—is a contour mill for a finish cut along the inner boundary face of the pocket using the same 0.75in. flat-end mill (T04). The last two operations drill the six holes, which consist of center drill using a 3/4×90DEG center drill bit (T18) and hole drilling using a 0.75×135° drill bit (T19). The toolpath of the hole drilling operations (including the center drill) is shown in Figure 5.5(e).

The pocket and holes are recognized as machinable features using the automatic feature recognition (AFR), and the machinable feature for the face milling operation will be created interactively. Like Lesson 4, we will use mostly the default options and parameter values prescribed in TechDB™.

You may open the example file with toolpaths created (filename: *2 point 5 axis feature with toolpath.SLDPRT*) to preview its toolpaths. When you open the file, you should see the six operations listed under the SOLIDWORKS CAM operation tree tab 🔲, as shown in Figure 5.6. You may simulate individual operations by right clicking the operation and choosing *Simulate Toolpath*. You may simulate combined operations by selecting them (press the control key) and pressing the right mouse button to select *Simulate Toolpath*. You may click *Mill Part Setup1* and choose the *Simulate Toolpath* button 🔘 Simulate Toolpath above the graphics area to see material removal simulation for all six operations combined.

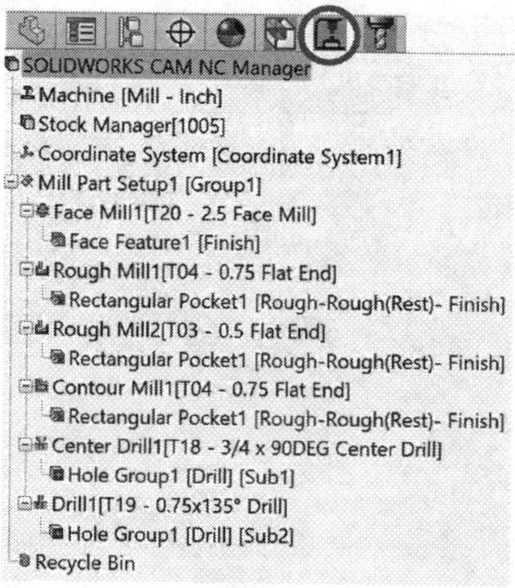

Figure 5.6 The six NC operations listed under the SOLIDWORKS CAM operation tree tab

5.3   Using SOLIDWORKS CAM

Open SOLIDWORKS Part

Open the part solid model (filename: *2 point 5 axis features.SLDPRT*) downloaded from the publisher's website. This solid model, as shown in Figure 5.2, consists of four solid features and a coordinate system. As soon as you open the model, you may want to check the unit system chosen and make sure the IPS system is selected. You may also increase the decimals from the default 2 to 4 digits following the steps similar to those discussed in Lesson 4.

Enter SOLIDWORKS CAM

Like what we learned in Lesson 4, at the top of the feature tree you should see two important tabs: SOLIDWORKS CAM feature tree 🔲 and SOLIDWORKS CAM operation tree 🔲 . You should also see SOLIDWORKS CAM command buttons in the ribbon bar above the graphics area (by clicking the SOLIDWORKS CAM tab above the graphics area; see Figure 1.9 of Lesson 1).

Select NC Machine

Click the SOLIDWORKS CAM feature tree tab ![img] and right click *Mill-inch* to select *Edit Definition*. Similar to those of Lesson 4, in the *Machine* dialog box, we select *Mill-inch* under *Machine* tab, choose *Tool Crib2* under *Available tool cribs* of the *Tool Crib* tab, select *M3AXIS-TUTORIAL* under the *Post Processor* tab, and select *Coordinate System1* under *Fixture Coordinate System* of the *Setup* tab.

Create Stock

From SOLIDWORKS CAM feature tree ![img], right click *Stock Manager* and choose *Edit Definition*. In the *Stock Manager* dialog box (Figure 5.7), we increase the height of the stock by 0.25in. on the top side (that is, enter *0.25* for *Z+*, circled in Figure 5.7). Choose *1005* for stock material. Accept the revised stock by clicking the checkmark ✓ at the top left corner of the dialog box. A rectangular stock should appear in the graphics window similar to that of Figure 5.4.

Mill Part Setup and Machinable Features

Click the *Extract Machinable Features* button ![img] above the graphics area (or choose from the pull-down menu *Tools > SOLIDWORKS CAM > Extract Machinable Features*). A *Mill Part Setup1* is created with two machinable features extracted: *Rectangular Pocket1* and *Hole Group1* (with six holes), all listed in SOLIDWORKS CAM feature tree ![img] (Figure 5.8). All machinable features are shown in magenta color since there are no NC operations generated for these features yet.

Generate Operation Plan and Toolpath

Click the *Generate Operation Plan* button ![img] above the graphics area. Five NC operations: 2 rough mills, 1 contour mill, 1 center drill, and 1 drill, are generated. They are listed in SOLIDWORKS CAM operation tree ![img], as shown in Figure 5.9.

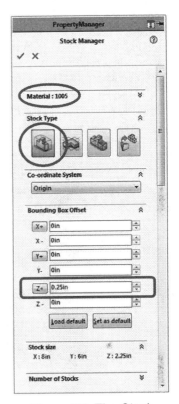

Figure 5.7 The *Stock Manager* dialog box

Figure 5.8 The machinable features extracted

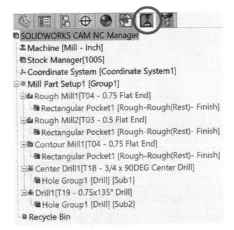

Figure 5.9 The operations generated

Again, these features are shown in magenta color, indicating that these operations are not completely defined yet. Click the *Generate Toolpath* button Generate Toolpath above the graphics area to create toolpath. The five operations are turned into a black color after toolpaths are generated.

Note that the default part setup origin coincides with the origin of the coordinate system (*Coordinate System1*), as circled in Figure 5.10, which is not desired for this example. As discussed in Lesson 4, the part setup origin defines the program zero location for the G-code. We relocate the origin to the front left corner at the top face of the stock. This corner point of the stock is easier to access at a CNC mill and is commonly chosen as the program zero in carrying out machining tasks. You may certainly select any adequate location as the G-code origin as long as the selected origin can be physically set up on the workbench or jig table of the mill at the shop floor.

The default part setup origin

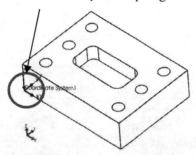

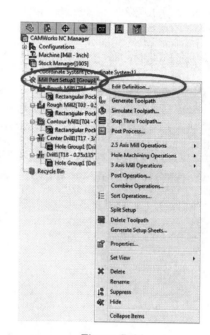

Figure 5.10 The default part setup origin
coinciding with *Coordinate System1*

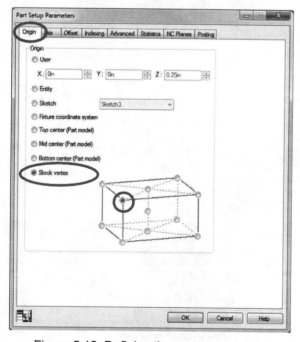

Figure 5.11

Part setup origin

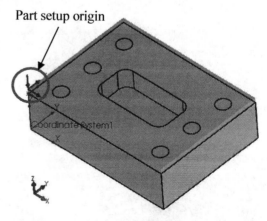

Figure 5.12 Defining the part setup origin

Figure 5.13 The part setup origin properly
relocated

Relocate the origin by right clicking *Mill Part Setup1* under the SOLIDWORKS CAM operation tree ◪ and select *Edit Definition* (see Figure 5.11).

In the *Part Setup Parameters* dialog box (Figure 5.12), choose *Stock vertex* (under the *Origin* tab), and pick the vertex at the front corner of the top face of the sample stock box, as circled in Figure 5.12. Click *OK* to accept the change. A warning box appears. Click *Yes* to the question in the warning box: *The origin or machining direction or advanced parameters has changed, toolpaths need to be recalculated. Regenerate toolpaths now?*

The toolpaths will be regenerated, and the part setup origin is now moved to the top left corner of the stock in the graphics area, as shown in Figure 5.13.

You may review individual operations listed under SOLIDWORKS CAM operation tree ◪ (see Figure 5.9), for example, *Rough Mill1*. Right click *Rough Mill1* and choose *Edit Definition*. In the *Operation Parameters* dialog box (Figure 5.14), *3/4 EM CRB 2FL 1-1/2 LOC* (that is, 0.75in. diameter flat-end mill, carbide, 2 flutes, and 1.5 in. length-of-cut) is chosen. Click the *Roughing* tab; the dialog box (Figure 5.15) shows *Pocket Out* as the pocketing pattern, stepover: 40% of tool diameter, and depth parameters: 50% of the tool diameter for both the *First cut amt.* and *Max cut amt.*

You may choose other tabs, such as *F/S* (feedrate and spindle speed), to review the machining parameters determined by the TechDB™. You may change these parameters and regenerate the toolpath as desired. We will simply click *Cancel* to close the dialog box.

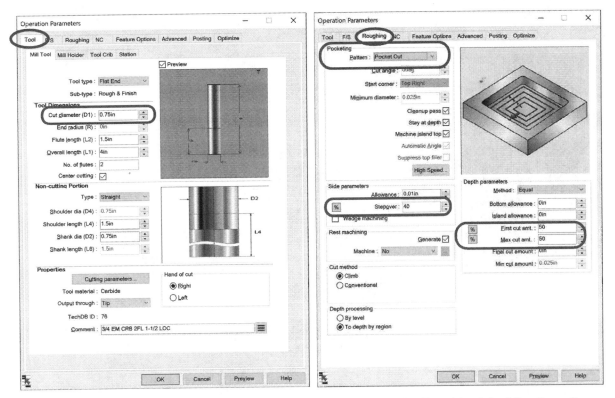

Figure 5.14 The *Mill Tool* tab of the *Operation Parameters* dialog box

Figure 5.15 The *Roughing* tab of the *Operation Parameters* dialog box

Simulate Toolpath

Click the *Simulate Toolpath* button ![Simulate Toolpath] above the graphics area. The *Toolpath Simulation* toolbox appears (like that of Figure 4.30 in Lesson 4). Click the *Run* button to simulate the toolpath. The machining simulation of all five operations will appear in the graphics area, similar to that of Figure 5.16.

Note that the face milling that removes the top 0.25in.-layer material of the stock has not been created. Pocket milling and hole drilling operations may not work well in practice without the face milling completed beforehand. We create a face milling operation next.

5.4   Creating a Face Milling Operation

We first tab the SOLIDWORKS CAM feature tree ![icon] to insert a 2.5 axis feature, and then follow the steps similar to those of Lesson 4 to create a machinable feature for face milling operation.

Manually Create Machinable Feature

Next we learn to manually create a machinable feature for the face milling operation.

From SOLIDWORKS CAM feature tree ![icon], right click *Mill Part Setup1* and choose *2.5 Axis Feature* (see Figure 5.17).

Figure 5.16 Material removal
simulation

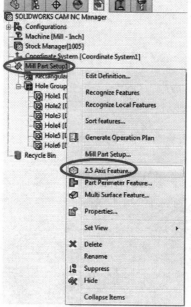

Figure 5.17 Create a new
2.5 axis feature

Figure 5.18 The *2.5 axis
Feature* dialog box

In the *2.5 Axis Feature* dialog box (Figure 5.18), choose *Face Feature* for *Type*, and pick the top face of the part in the graphics area (see Figure 5.19). The top face is highlighted. The selected face (*CW Face-1*) is now listed under *Selected Entities* in the *2.5 Axis Feature* dialog box (circled in Figure 5.18).

Click the *Next* button 🔘 (circled in Figure 5.18) to define the end condition. Choose *Finish* for *Strategy* (see Figure 5.20), and choose *Upto stock* for the *End condition: Direction 1*. Click the *Reverse direction* button 🔁 if necessary. Make sure *0.25in* appears for dimension, as circled in Figure 5.20. Click the checkmark ✓ to accept the machinable feature.

A *Face Feature1* node is now listed in the SOLIDWORKS CAM feature tree 🔲 in magenta color (see Figure 5.21). Right click *Face Feature1* and choose *Generate Operation Plan*. One new node, *Face Mill1*, is now listed in SOLIDWORKS CAM operation tree 🔲 (Figure 5.22). Right click the node and choose *Generate Toolpath*. A face milling toolpath with a 2in. face mill cutter is generated like that shown in Figure 5.23.

Pick the top face

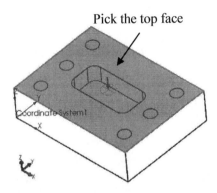

Figure 5.19 Picking the top face of the part to create a machinable feature for the face milling operation

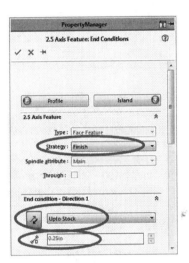

Figure 5.20 Defining end conditions

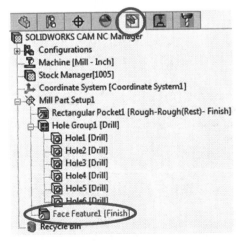

Figure 5.21 A *Face Feature1* machinable feature created

Figure 5.22 A *Face Mill1* operation generated

Choose a Different Cutter

We modify the face milling operation by choosing a 2.5in. face mill (just to learn how to add a cutter to the tool crib) and enter 0.125in. for depth of cut.

Right click *Face Mill1* in the SOLIDWORKS CAM operation tree ▣ and choose *Edit Definition*. In the *Operation Parameters* dialog box, choose *the Tool* tab; the 2in. face mill is shown (similar to Figure 5.24). Since the 2.5in. face mill is not in the current tool crib, we will need to add it to the crib.

First, we choose the *Tool Crib* tab and click the checkbox in front of the *Filter* button to display only the face mill cutters (see Figure 5.25).

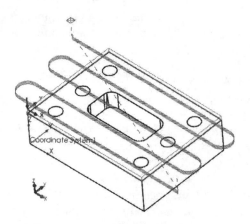

Figure 5.23 The toolpath of *Face Mill1* operation

Click the *Filter* button. In the *Tool Select Filter* dialog box (Figure 5.26), choose *Face Mill* for *Type*, and click the *Filter by* checkbox, then click *OK*. In the *Operation Parameters* dialog box, only one face mill is listed.

Click the *Add* button (see Figure 5.27) to add a face mill tool. In the *Tool Select Filter* dialog box (Figure 5.28), select *Face Mill* for *Tool type*, click the tool in the third row (ID:3) to select the 2.5in. face mill, and then click *OK*.

A 2.5in. face mill is now listed under *Tools* in the *Operation Parameters* dialog box (see Figure 5.29). Choose the 2.5in. face mill and click *Select*. Click *Yes* to the question: *Do you want to replace the corresponding holder also?* We have now replaced the cutter with a 2.5in. face mill. We will modify the depth parameters next.

Choose the *Facing* tab of the *Operation Parameters* dialog box (see Figure 5.30). In the *Depth parameters* group, click the percentage button ▣ to deselect it for both the *First cut amt.* and *Max cut amt.* Enter *0.125in.* for both *First cut amt.* and *Max cut amt.*, as shown in Figure 5.30. Click *OK* to accept the changes. The face milling toolpath will be generated like that shown in Figure 5.5(a) with two rounds of passes, cutting 0.125in. depth in the first round and then the remaining 0.125in. in the next round to finish it up.

Simulate Toolpath

Click the *Simulate Toolpath* button 📷 Simulate Toolpath above the graphics area. The *Toolpath Simulation* toolbox appears (see Figure 4.30 of Lesson 4). Click the *Run* button ▶ to simulate the toolpath. The material removal simulation of all six operations will appear in the graphics area at the end, similar to that of Figure 5.1.

5.5 Re-Ordering Machining Operations

Now we have created all six operations. However, their order is off. We would like to see these six operations in order as follows: *Face Mill1*, *Rough Mill1*, *Rough Mill2*, *Contour Mill1*, *Center Drill1*, and *Drill1*, similar to that shown in Figure 5.6. To reorder these operations, you may click and drag them one at a time (or by pressing the shift key to select multiple) and move them up or down in the operation tree.

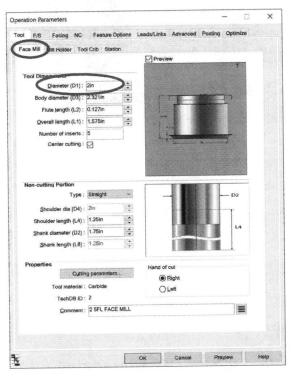

Figure 5.24 The *Face Mill* tab of the *Operation Parameters* dialog box

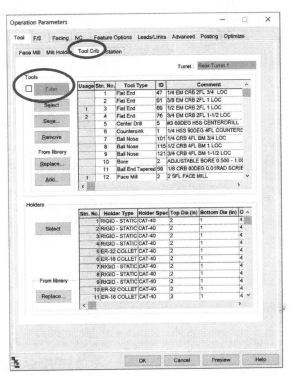

Figure 5.25 Displaying only the face mill cutters using the filter option under the *Tool Crib* tab

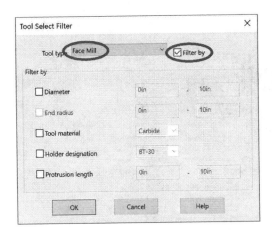

Figure 5.26 The *Tool Select Filter* dialog box

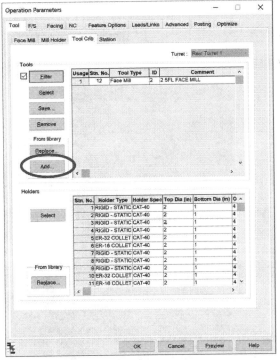

Figure 5.27 Clicking the *Add* button to add face mill under the *Tool Crib* tab

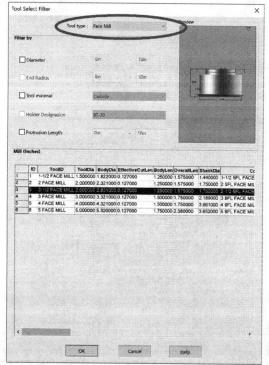

Figure 5.28 Choosing a 2.5in. face mill

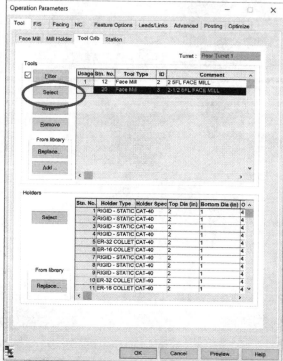

Figure 5.29 Selecting the 2.5in. face mill

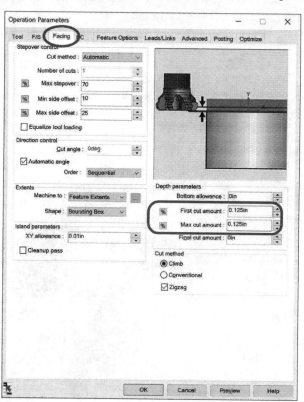

Figure 5.30 Defining depth parameters under the *Facing* tab

In this case, all you have to do is to click and drag *FaceMill1* and place it above *RoughMill1*.

You may click the *Simulate Toolpath* button _{Toolpath} above the graphics area to review material removal simulation for all six operations combined in a desired order.

5.6   Reviewing Machining Time

You may choose to review machining time for individual operations or the overall time for combined operations.

From SOLIDWORKS CAM operation tree 🔲 , right click an operation, for example *Face Mill1*, and choose *Edit Definition*.

In the *Operation Parameters* dialog box (Figure 5.31), choose *Optimize* tab, and look for *Estimated machining time*. The estimated feed time of *Face Mill1* operation is 2.544 minutes.

Note that you may choose *Mill Part Setup1* and right click *Edit Definition* to review the machining time for the combined six operations. Choose *Statistics* tab in the *Part Setup Parameters* dialog box (Figure 5.32). The overall machining time for all six operations is 50.65 minutes. You may calculate the toolpath length by sketching the toolpath (similar to that of Lesson 4) to verify the machining time.

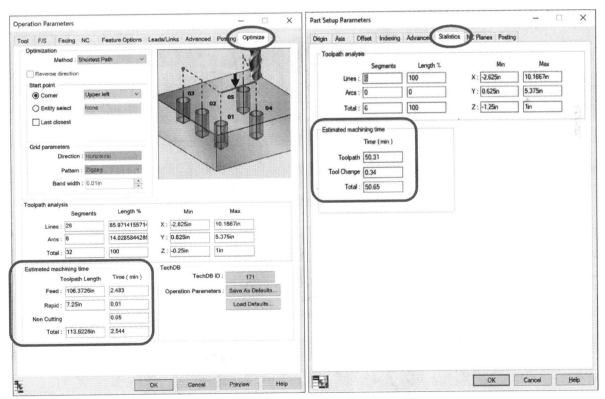

Figure 5.31 Machining time of the *Face Mill1* operation

Figure 5.32 Machining time of the combined six NC operations

```
O0001                          N63 X2.985                    N177 G00 Z-.15                 N332 G90 G54 G00 X1. Y5.
N1 G20                         N64 Y3.615                    N178 Z-.9                      N333 G43 Z1. H17 M08
N2 (2-1/2 6FL FACE MILL)       N65 X2.5                      N179 G01 Z-1.25 F1.4381        N334 G82 G98 R-.15 Z-.5875 P00
N3 G91 G28 X0 Y0 Z0            N66 G03 X2.385 Y3.5 I0 J-.115 ...                            F3.3965
N4 T20 M06                     N67 G01 Y2.5                                                 N335 Y3.
N5 S693 M03                    N68 G03 X2.5 Y2.385 I.115 J0  N274 ( Contour Mill1 )         N336 Y1.
                               N69 G01 X5.5                  N275 G90 G54 G00 X4.3663 Y3.3273  N337 X7.
N6 ( Face Mill1 )              ...                           N276 G43 Z-.15 H04 M08         N338 Y3.
N7 G90 G54 G00 X-2.625 Y5.375                                N277 G01 Z-.625 F.8556         N339 Y5.
N8 G43 Z1. H20 M08             N155 ( Rough Mill2 )          N278 G41 D24 X4.0905 Y3.603 F2.5669  N340 G80 Z1. M09
N9 G01 Z-.125 F5.              N156 G90 G54 G00 X2.5 Y3.74   N279 G03 X4.0375 Y3.625 I-.053 J-.053  N341 G91 G28 Z0
N10 G17 X0 F34.3362           N157 G43 Z-.15 H03 M08        N280 G01 X2.5 F3.4225          N342 (3/4 SCREW MACH DRILL)
N11 X8. F45.7816              N158 G01 Z-.5 F1.4381         N281 G03 X2.375 Y3.5 I0 J-.125  N343 T18 M06
N12 X9.375                    N159 G03 X2.26 Y3.5 I0 J-.24 F5.7525  N282 G01 Y2.5          N344 S595 M03
N13 G02 Y3.7917 I0 J-.7917    N160 G01 Y3.48                N283 G03 X2.5 Y2.375 I.125 J0
N14 G01 X8.                   N161 G02 X2.52 Y3.74 I.26 J0  N284 G01 X5.5                  N345 ( Drill1 )
....                          N162 G01 X2.5                 N285 G03 X5.625 Y2.5 I0 J.125  N346 G90 G54 G00 X1. Y5.
                              N163 G00 Z-.15                N286 G01 Y3.5                  N347 G43 Z1. H18 M08
N50 ( Rough Mill1 )           N164 Z-.4                     N287 G03 X5.5 Y3.625 I-.125 J0  N348 G83 G98 R-.15 Z-1.25 Q.1
N51 G90 G54 G00 X2.985 Y3.015  N165 G01 Z-.75 F1.4381       N288 G01 X3.9625              F3.3965
N52 G43 Z-.15 H04 M08         N166 G03 X2.26 Y3.5 I0 J-.24 F5.7525  N289 G03 X3.9095 Y3.603 I0 J-.075  N349 Y3.
N53 G01 Z-.625 F.8556         N167 G01 Y3.48                N290 G40 G01 X3.6337 Y3.3273   N350 Y1.
N54 Y2.985 F3.4225            N168 G02 X2.52 Y3.74 I.26 J0  N291 G00 Z-.15                 N351 X7.
N55 X5.015                    N169 G01 X2.5                 N292 X4.3663                   N352 Y3.
N56 Y3.015                    N170 G00 Z-.15                N293 Z-.525                    N353 Y5.
N57 X2.985                    N171 Z-.65                    N294 G01 Z-.9375 F.8556        N354 G80 Z1. M09
N58 Y3.315                    N172 G01 Z-1. F1.4381         N295 G41 D24 X4.0905 Y3.603 F2.5669  N355 G91 G28 Z0
N59 X2.685                    N173 G03 X2.26 Y3.5 I0 J-.24 F5.7525  N296 G03 X4.0375 Y3.625 I-.053 J-.053  N356 G28 X0 Y0
N60 Y2.685                    N174 G01 Y3.48                ...                            N357 M30
N61 X5.315                    N175 G02 X2.52 Y3.74 I.26 J0
N62 Y3.315                    N176 G01 X2.5                 N331 ( Center Drill1 )
```

Figure 5.33 Partial contents of the .txt file (G-code)

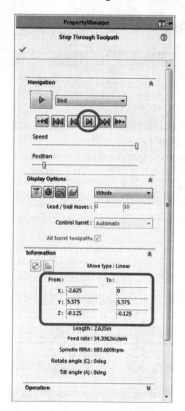

Figure 5.34 The *Step
Through Toolpath* dialog box

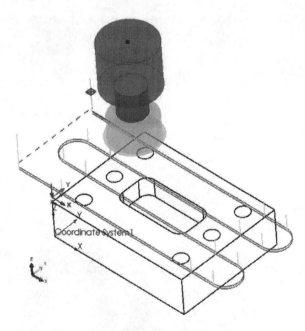

Figure 5.35 The cutter and
toolpath displayed in the graphics

5.7   The Post Process and G-code

You may click the *Post Process* button ▣ ₚₒₛₜ above the graphics area, and follow the same steps learned in Lesson 4 to convert the toolpaths into G-code. Figure 5.33 shows partial contents of the G-code file. Note that similar to what we discussed in Lesson 4, the NC Words, G41 D24 (for examples in Blocks N278 and N295), are not needed.

We will take a closer look at some of the NC blocks next.

5.8   Stepping Through the Toolpath

We first take a closer look at the face milling toolpath by right clicking it (under the SOLIDWORKS CAM operation tree ▣) and choosing *Step Through Toolpath*. In the *Step Through Toolpath* dialog box (Figure 5.34), click the *Forward single step* button ▣ six times to move the cutter to X: 0, Y: 5.375, and Z: –0.125 position. The cutter is in contact with the stock, as shown in Figure 5.33. This cutter location takes place after the NC block N10, in which X is 0. Z and Y positions are determined by blocks N9 and N7, respectively (see Figure 5.33). The next block (N11) moves the cutter to the front end of the stock, where X is 8, Y- and Z-coordinates stay the same (Y = 5.375, Z = –0.125). Note that these X-, Y-, and Z-coordinates refer to the part setup origin (again located at the front left corner of the top face of the stock), as they should be.

You may select other operations and step through the toolpaths. You will find the cutter locations (see X, Y, Z shown in the *Step Through Toolpath* dialog box, circled in Figure 5.34) are consistent with those in the G-code.

We have completed this tutorial lesson. You may save your model for future reference.

5.9   Exercises

Problem 5.1. Generate machining operations to cut the part shown in Figure 5.36 from a rectangular stock of 4in.×3in.×1.25in. Use cutters and machining parameters chosen by SOLIDWORKS CAM. Please submit the following for grading:

(a)   A summary of the NC operations, including number of operations, cutting parameters, and tools selected.
(b)   Screen shots of combined NC toolpaths and material removal simulations.

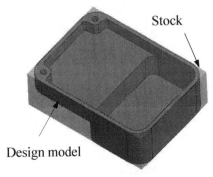

Figure 5.36 The design model and stock of Problem 5.1

Note that you will have to create two mill part setups to cut the features from the top and bottom of the part, respectively.

Problem 5.2. Use SOLIDWORKS CAM to generate NC sequences to cut the part shown in Figure 5.37 from a rectangular block of 4in.×4.25in.×1.55in. (material: steel 1005). Pick your own cutters and choose/enter adequate machining parameters with justifications. Please submit the following for grading:

(a)   A summary of the NC operations, including number of operations, cutting parameters, and tools selected.

(b)   Screen shots of combined NC toolpaths and material removal simulations.

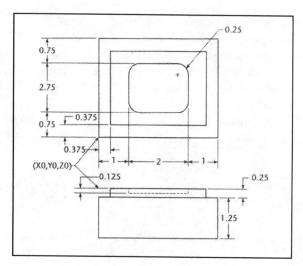

Figure 5.37 The design model of Problem 5.2

Lesson 6: Machining a Freeform Surface and Limitations

6.1   Overview of the Lesson

In this lesson, we focus on creating machining simulation for cutting a freeform surface (also called contoured surface or sculptural surface) often seen in mold or die machining.

In general, machining a die or mold with a freeform cavity surface requires a rough cut (often referred to as volume milling) that removes material from a raw stock as fast as possible using a larger tool, and therefore, a larger stepover and depth of cut. Since a larger tool and steps are employed, more material remains on the freeform surface uncut at the end of a rough cut. Stock at this stage may not be ready for a finish cut that polishes the stock-in-progress with desired accuracy and surface finish quality. Therefore, in practice we often insert a local milling in between, in which a smaller tool with smaller stepover and depth of cut are employed to remove the material remaining on the surface after the volume milling. A local milling removes the noticeable amount of material remaining on the surface and prepares the stock-in-progress properly for the final finish cuts. Both volume milling and local milling are cutting the stock following a prescribed cut pattern (such as zigzag, pocket out, etc.) slice by slice. The cutter moves on the X-Y plane while Z-coordinate is set to the prescribed slice thickness, which is determined by the depth of cut. The surface of the stock-in-progress after volume milling or combined volume milling and local milling operations is staircase-like since the material is removed slice by slice. The final finish cut is often carried out by a surface milling (3-, 4-, or 5-axis), in which the toolpath lays right on the freeform surface, and the cutter is directly in contact with the freeform surface, in which cutter is moving in all axes simultaneously. With a reasonably small stepover, the surface milling operation is able to produce a finished surface that meets the requirement of surface finish, which is often characterized by the scallop height in the context of machining. Note that often a second (or more) surface milling operation is needed to further polish the sculptural surface to meet the accuracy and finish requirements.

In SOLIDWORKS CAM, a default machining strategy, *Area Clearance, Z Level*, is assigned to cut a freeform surface. This strategy leads to two machining operations, *Area Clearance* and *Z Level*. The *Area Clearance* machining operation performs a rough cut (or volume milling). This operation first slices the material to be removed from the raw stock into layers that are normal to the Z-axis. Again, the thickness of the layer is defined by depth of cut (or cut amount). A scan (or cut) pattern, for example, zigzag, can be chosen to move the cutter on the X-Y plane in cutting the material slice by slice. The *Z Level* operation performs a surface milling that removes material by making a series of horizontal planar cuts. The cuts follow the contour of the freeform surface at decreasing Z levels. That is, at each level, the Z-coordinate is fixed and the cutter moves in X- and Y-axes simultaneously. Therefore, *Z Level* operation is not a contour surface milling operation, in which cutter moves in all three axes simultaneously. Usually, a smaller cutter and a smaller depth of cut are employed for the *Z Level* operation in attempt for a better surface finish. The surface of the stock after *Z Level* operation is still staircase-like.

One major limitation of SOLIDWORKS CAM 2023 in cutting a freeform surface is that SOLIDWORKS CAM does not offer a surface milling operation that supports finish cut, although such a machining operation is commonly found in other CAM software. For example, the *Pattern Project* strategy of CAMWorks, *Multi-Axis Milling* of HSMWorks, and *Multi-Axis Toolpath* of Mastercam all support surface milling operations that perform desired finish cut.

In this lesson we will machine a freeform surface—see Figure 6.1(a)—using the default *Area Clearance, Z Level* machining strategy. Since a freeform surface is not a standard 2.5 axis feature that AFR is able to extract automatically, we will learn how to manually select surfaces and create a machinable feature. We will then generate an operation plan, generate a toolpath, and simulate a toolpath, similar to those we learned in previous lessons. We will carry out material removal simulation like that of Figure 6.1(b). We take a closer look at the operations generated by SOLIDWORKS CAM using default settings, including the options and parameters determined by TechDB™. We then modify the *Area Clearance* operation by adjusting machining parameter (such as stepover and depth of cut) and selecting a different size cutter in hope to remove the material faster. Thereafter, we adjust the cut amount for the *Z Level* operation to see if a freeform surface can be cut to meet a desired surface finish since the default settings do not provide a desired cut. Note that due to a large curvature variation of the freeform surface in the solid model, even a ball-nose cutter is not able to reach some areas for an accurate cut—circled in Figure 6.1(c), following the toolpaths generated by SOLIDWORKS CAM.

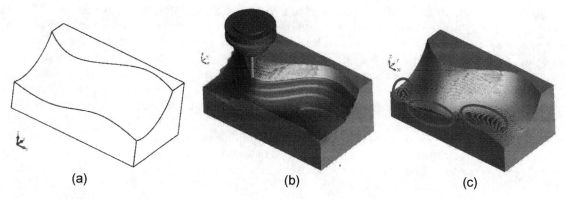

(a)   (b)   (c)

Figure 6.1 The freeform surface example, (a) part solid model, (b) material removal simulation, and (c) excessive amount of material remaining on the machined surface

In Lesson 14, we revisit this example by using a multiaxis surface milling operation offered in CAM modules integrated with SOLIDWORKS, including CAMWorks, HSMWork, and Mastercam to SOLIDWORKS. A multiaxis surface milling operation will be able to clean up the material remaining on the freeform surface and create a desirable surface finish.

6.2   The Freeform Surface Example

The size of the bounding box of the part (filename: *Freeform Surface.SLDPRT*) is 7.5in.×4in.×3in. The solid model is lofted (or blend) across four parallel sketches, each spaced 2.5in. apart, as shown in Figure 6.2(a). Similarly, the freeform surface is formed by lofting the circular arcs of respective sketches along the longitudinal direction (X-direction of *Coordinate System1*). Individual sketches are created by four straight lines and a circular arc. The front and rear end faces of the loft share the same sketch (that is, Sketch1 and Sketch4 shown in Figure 6.2(a) are identical). Dimensions of individual sketches are shown in Figure 6.2(b), (c), and (d), respectively.

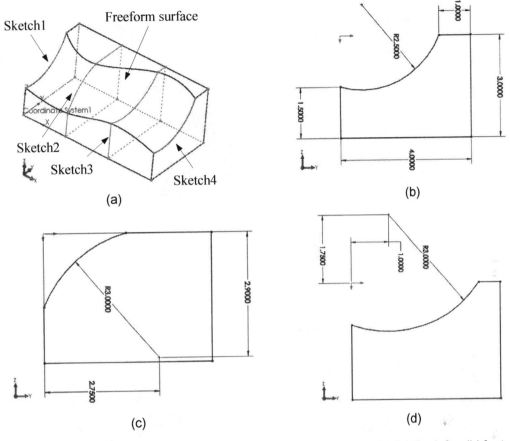

Figure 6.2 Sketches and dimensions of the freeform surface example: (a) the loft solid feature, (b) Sketch 1 (and 4), (3) Sketch 2, and (d) Sketch 3

In addition, a fixture coordinate system (*Coordinate System1*) is defined at the front left corner of the bottom face of the part. The unit system chosen is IPS (inch, pound, second). When you open the solid model *Freeform Surface.SLDPRT*, you should see the loft solid feature and a coordinate system listed in the feature tree like that of Figure 6.3.

A stock of rectangular block with a size 7.5in.×4in.×3in., made of low carbon alloy steel (1005), as shown in Figure 6.4, is chosen for the machining operations. Note that a part setup origin is defined at the front left vertex on the top face of the stock, which locates the G-code program zero.

As mentioned above, we implement two machining operations for this example. The first operation is *Area Clearance1*, which is a rough cut using a 0.75in. hog nose mill with 0.125in. corner radius (T19 in tool crib). The toolpath of the *Area Clearance1* operation is shown in Figure 6.5(a). The second operation is a local milling (*Z Level1*) that continues removing the material remaining from *Area Clearance1* using a smaller ball nose cutter of diameter 0.5in. (T07); see toolpath in Figure 6.5(b).

You may open the example file with toolpath created (filename: *Freeform Surface with toolpath.SLDPRT*) to preview the toolpath in this example. When you open the file, you should see the two operations listed under the SOLIDWORKS CAM operation tree tab ▣ , as shown in Figure 6.6. You may simulate individual operations by right clicking it and choosing *Simulate Toolpath*, or simulate the combined operations by clicking the *Simulate Toolpath* button above the graphics area.

Figure 6.3 Feature tree of the
freeform surface example

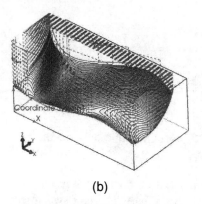

Figure 6.4 Stock with a part setup origin
created at the front left vertex on top face

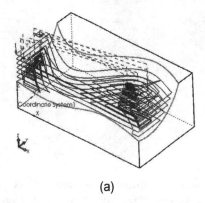

(a)

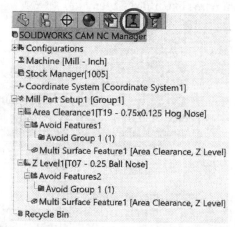

(b)

Figure 6.5 Toolpaths of the two NC operations: (a) *Area Clearance1*, and (b) local milling (*Z Level*)

6.3 Using SOLIDWORKS CAM

Open SOLIDWORKS Part

Open the part file (filename: *Freeform Surface.SLDPRT*) downloaded from the publisher's website. This solid model, as shown in Figure 6.2(a), consists of one loft solid feature and a coordinate system. As soon as you open the model, you may want to check the unit system chosen and make sure the IPS system is selected. You may also increase the decimals from the default 2 to 4 digits similar to that of previous lessons.

Figure 6.6 The NC operations listed under
the SOLIDWORKS CAM operation tree tab

Select NC Machine

Click the SOLIDWORKS CAM feature tree tab 📚 and right click *Machine [Mill-inch]* to select *Edit Definition*. Similar to the previous lessons, in the *Machine* dialog box, we select *Mill-inch* under *Machine* tab, choose *Tool Crib2* under *Available tool cribs* of the *Tool Crib* tab, select *M3AXIS-TUTORIAL* under the *Post Processor* tab, and select *Coordinate System1* under *Fixture Coordinate System* of the *Setup* tab.

Create Stock

From SOLIDWORKS CAM feature tree tab 📚, right click *Stock Manager* and choose *Edit Definition*. In the *Stock Manager* dialog box, we choose the default stock size (7.5in.×4in.×3in.) and material (*Steel 1005*). The rectangular stock should appear in the graphics window similar to that of Figure 6.4.

We will first select the freeform surface to define a machinable feature and select *Area Clearance, Z Level* machining strategy. We will let SOLIDWORKS CAM technology database determine machining settings, and then modify them for a better machined surface with reduced machining time. We will relocate the part setup origin to the front left corner at the top face of the stock, as shown in Figure 6.4.

Create a Machinable Feature

Click the SOLIDWORKS CAM feature tree tab 📚, right click *Stock Manager*, and choose *Mill Part Setup*, as shown in Figure 6.7.

The *Mill Setup* dialog box appears and the *Front Plane*, perpendicular to the Z-axis, is selected (if not, select *Front Plane*). We choose *Front Plane* to define tool axis (or feed direction) of the setup. The *Front Plane* should appear under *Entity* in the dialog box. In the graphics area (see Figure 6.8), a symbol 🔾 with arrow pointing upward appears. This symbol indicates that the tool axis must be reversed (pointing downward). Click the *Reverse Selected Entity* button 🔲 under *Entity* to reverse the direction. Make sure that the arrow points in a downward direction. Click the checkmark ✔ to accept the definition. A *Mill Part Setup1* is now listed in the feature tree.

Now we create a machinable feature. From SOLIDWORKS CAM feature tree 📚, right click *Mill Part Setup1* and choose *Multi Surface Feature* (see Figure 6.9).

In the *Multi Surface Feature* dialog box (Figure 6.10), pick the freeform surface of the part in the graphics area; the surface picked is now listed under *Selected Faces*. Leave the default *Area Clearance, Z Level* for *Strategy*, and click the checkmark ✔ to accept the surface feature.

Figure 6.7 Selecting *Mill Part Setup*

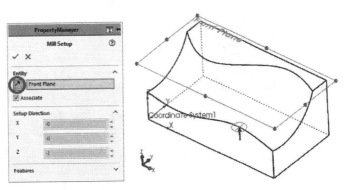

Figure 6.8 Picking the *Top Plane* for mill setup

Note that the other two machining strategies available are *Coarse* and *Finish*. SOLIDWORKS CAM indicates that no operations have been defined in the technology database if either one is selected. Therefore, the default strategy *Area Clearance, Z Level* is the only option available.

In the SOLIDWORKS CAM feature tree 🔖, a *Multi Surface Feature1* is added in magenta color under *Mill Part Setup1*, as shown in Figure 6.11.

Next, we create another multi surface feature and pick the top face of the part to define it as the area to avoid. This is to restrict the toolpath to stay only on the freeform surface.

Right click *Mill Part Setup1* and choose *Multi Surface Feature*. In the *Multi Surface Feature* dialog box, pick the top face of the part in the graphics area (see Figure 6.12), click the *Define as Avoid Feature* box, and click the checkmark ✔ to accept it.

In the SOLIDWORKS CAM feature tree 🔖, a *Multi Surface Feature2[Avoid]* is added in magenta color under *Mill Part Setup1*.

Generate Operation Plan and Toolpath

Click the *Generate Operation Plan* button 🔲 above the graphics area. Two operations, *Area Clearance1* and *Z Level1*, are generated. They are listed in SOLIDWORKS CAM operation tree 🔲 (as seen in Figure 6.13). Again, they are shown in magenta color, indicating that these operations are not completely defined yet. If you expand an operation (for example, *Area Clearance1*) in the feature tree, an *Avoid Feature1* node is listed. Expand the *Avoid Feature*; *Avoid Group 1 (0)* appears, indicating that the feature to avoid has not been selected ("0" appears in the parentheses). We will accept the operations as they are for now. We will come back momentarily to revisit the avoid feature selection.

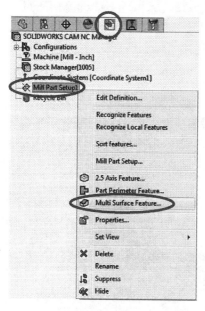

Figure 6.9 Choosing *Multi Surface Feature*

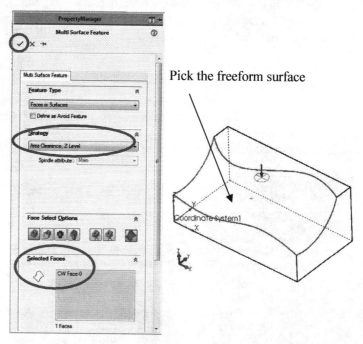

Pick the freeform surface

Figure 6.10 Picking the freeform surface for machinable feature

Note that when you click the *Mill Part Setup* node (under SOLIDWORKS CAM operation tree 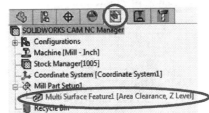), the default part setup origin appears at the front left corner of the bottom face of the stock, coinciding with the coordinate system circled in Figure 6.14, which is less desirable. We will relocate the origin to the front left corner at the top face of the stock.

Redefine the part setup origin by right clicking *Mill Part Setup1* in the SOLIDWORKS CAM operation tree 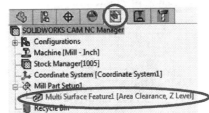 and selecting *Edit Definiton*. In the *Part Setup Parameters* dialog box (see Figure 6.15), choose *Stock vertex* (under the *Origin* tab), and pick the vertex at the front left corner of the top face. Click *OK* to accept the change. The part setup origin is now moved to the front left corner of the top face of the stock in the graphics area like that of Figure 6.4.

Figure 6.11

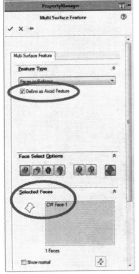

Pick the top face to avoid

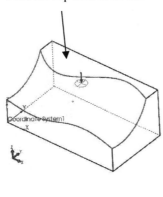

Figure 6.12 Picking the top face as a face to avoid

The current part setup origin

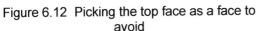

Figure 6.13 The two operations generated

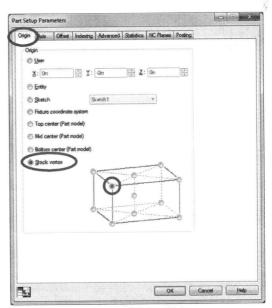

Figure 6.14 The default part setup origin located at the center of the stock top face

Figure 6.15 Selecting vertex in the *Part Setup Parameters* dialog box

Click the *Generate Toolpath* button ⟨Generate Toolpath⟩ above the graphics area to create the toolpath. The two operations are turned into black color right after toolpaths are generated.

Toolpaths of the two operations, *Area Clearance* and *Z Level*, are generated like those in Figure 6.16(a) and Figure 6.17(a), respectively. The toolpath of *Area Clearance* appears to not only remove material above the freeform surface, but also to cut the area below the top face of the part (behind the freeform surface).

The material removal simulations, shown in Figure 6.16(b), indicate the cutter plunged into the stock material behind the freeform surface. Furthermore, at the end of the material removal simulation carried out for the combined operations, Figure 6.17(b) reveals that excessive material of staircase like appearance remained uncut on the freeform surface, leading to undesirable surface finish.

These toolpaths are certainly not desirable. We need to at least contain the toolpaths to only machine the freeform surface by selecting the top face of the part as the avoid feature for the *Area Clearance* operation. We will also make an attempt to improve the surface finish by adjusting machining parameters of the *Z Level* operation.

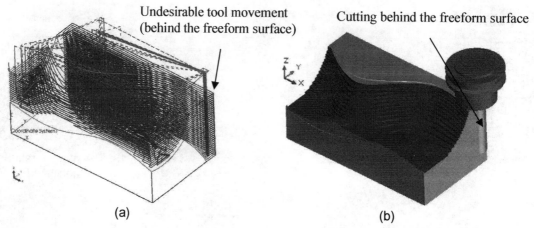

Figure 6.16 The *Area Clearance1* operation, (a) toolpath, and (b) material removal simulation

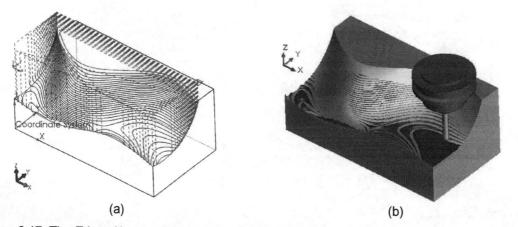

Figure 6.17 The *Z Level1* operation, (a) toolpath, and (b) material removal simulation (combined with *Area Clearance* operation)

6.4   Selecting Avoid Feature to Correct the Toolpath

Right click *Avoid Group 1(0)* under *Avoid Features1* of *Area Clearance1* and select *Edit Definition* (see Figure 6.18).

Click the checkbox in front of the *Multi Surface Feature2[Avoid]* in the *Avoid Features* dialog box (Figure 6.19), then click *OK*. The number in the parentheses of *Avoid Group 1* under the feature tree should become *1* (was *0*).

Repeat the same steps for the *Z Level1* operation (just to make sure).

Click the *Generate Toolpath* button above the graphics area. The toolpaths will be regenerated like those shown in Figure 6.20. Note that the tool movements of *Area Clearance* operation behind the freeform surface have been eliminated as desired.

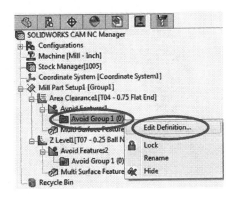

Figure 6.18

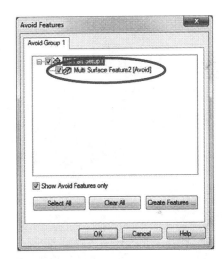

Figure 6.19 The *Avoid Features* dialog box

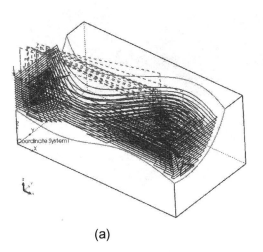

(a)

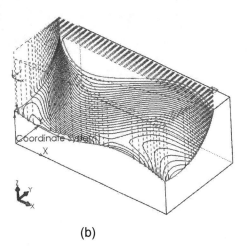

(b)

Figure 6.20 Toolpaths with avoid features selected, (a) *Area Clearance1*, and (b) *Z Level1*

Simulate the toolpaths of combined operations (click the *Simulate Toolpath* button 🔧 above the graphics area). The material removal simulations shown in Figure 6.21(a) and Figure 6.21(b) indicate that not only is the surface unsmooth but material remained uncut on the freeform surface. How much is the material remaining uncut? We take a closer look next.

In the *Toolpath Simulation* tool box, select the *Show difference* button 🔲 under *Display Options*—circled in Figure 6.22(a)—to show a visual comparison of the machined part and the design part in the graphics area.

The result is shown in Figure 6.22(b), which indicates the cut is indeed not clean. For example, the area close to the middle of the front edge of the freeform surface circled in Figure 6.22(b) shows dark blue color indicating an amount of more than 0.0157in. undercut material, as depicted in the color spectrum to the left. Other areas, such as those close to the right and left sides of the freeform surface circled in Figure 6.22(b), show a similar issue. Staircase type scallops remain on the surface.

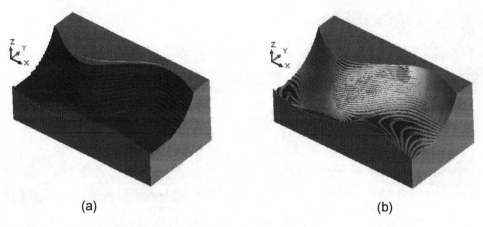

(a)   (b)

Figure 6.21 The material removal simulation, (a) *Area Clearance1* operation only, and (b) combined operations, including *Z Level1* operation

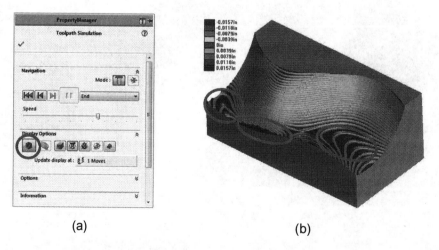

(a)   (b)

Figure 6.22 The material removal simulation, (a) the *Show Difference* button of the *Toolpath Simulation* tool box, and (b) differences between the machined part and the design part shown with a color spectrum

How long does the entire machining operation take? The machining times are estimated as 272 and 7.5 minutes, respectively, for *Area Clearance1* and *Z Level1* operations, based on the feedrates determined by the TechDB™.

Can the machining time of the *Area Clearance1* operation be reduced (by using the feedrates determined by the TechDB™)? Is the toolpath of *Z Level* fine enough to produce a satisfactory finished surface? Can we modify the toolpath of the rough cut (*Area Clearance*) with reduced machining time and improve the toolpath for a better surface finish (*Z Level*) with better accuracy? Furthermore, we did find the areas that show significant deviation between the finished part and the design part in Figure 6.22(b). Can this deviation be reduced or completely eliminated?

6.5   Modifying the Area Clearance Toolpath

How do we modify the toolpath? Where to find and modify machining parameters that may adjust the toolpath, such as depth of cut, stepover, and cut pattern? First, we take a closer look at the first operation, *Area Clearance1*. We would like to use a larger cutter and larger stepover and depth of cut to reduce the machining time.

In SOLIDWORKS CAM operation tree ⬛ , right click *Area Clearance1* and choose *Edit Definition*. In the *Operation Parameters* dialog box, Figure 6.23(a), T76 (tool ID designated in TechDB™) has been selected (0.75in. flat-end mill) under the *Tool* tab. Ideally, we would like to use a large-size cutter for this operation. However, there is no larger cutter that is suitable for this operation in the tool library. Hence, we will use a 0.75in. (same outer diameter) hog nose cutter with a corner radius 0.125in. for this operation, just to show you how to add and choose a different cutter.

In the *Operation Parameters* dialog box, choose *Tool Crib* tab. Similar to that of Lesson 5, we use the *Filter* option—see Figure 6.23(b) and Figure 6.23(c)—to display hog nose cutters only. Since there is no hog nose cutter in the tool crib, we will have to add one from the tool library we intend to use.

Click *Add* under the *Tool Crib* tab; see Figure 6.23(b). In the *Tool Select Filter* dialog box, select tool ID 170 (3/4 CRB 4FL HGN .125R 1-1/2 LOC) and click *OK*; see Figure 6.23(d). The tool will be listed under the *Tool Crib* tab; see Figure 6.23(e). Click the cutter and click *Select*. Click *Yes* to the question: *Do you want to replace the corresponding holder also?* Choose *Mill Tool* tab to review details of the hog nose cutter selected; see Figure 6.23(f).

We have now replaced the tool with a modified 0.75in. hog nose cutter with 0.125in. corner radius. We will modify the machining parameters next.

Choose the *Pattern* tab of the *Operation Parameters* dialog box, select *Pocket Out* for *Pattern*, and enter 80% and 40% for *Max.* and *Min. stepover*, respectively, as shown in Figure 6.24(a).

Choose the *Area Clearance* tab of the *Operation Parameters* dialog box; see Figure 6.24(b). In the *Depth* parameters group, enter *0.25in.* for *Cut amount*, as shown in Figure 6.24(b). Note that the cut amount 0.25in. is 1/3 of the cutter diameter, which is generally considered adequate in practice. Click *OK* to accept the changes and click *Yes* to the warning message: *Operation parameters have changed, toolpaths need to be recalculated. Regenerate toolpaths now?*

The area clearance toolpath will be regenerated to be like that shown in Figure 6.5(a). The machining time is now 75.0 minutes (click the *Statistics* tab to see the machining time estimated), as shown in Figure 6.24(c), based on the feedrate, 4.930 in./min. determined by the TechDB™, stated under the *F/S* tab.

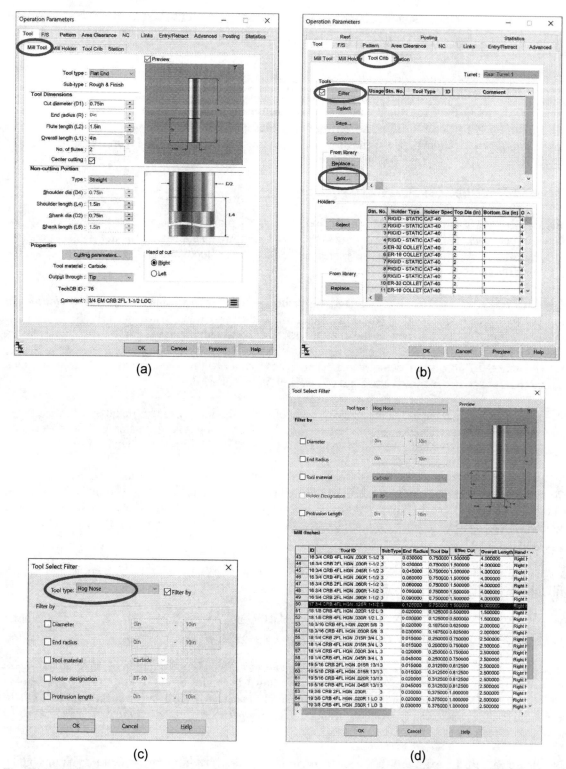

Figure 6.23 Replacing the tool with a 0.75in. hog nose, (a) 0.75in. flat-end mill selected currently, (b) showing no hog nose cutter in the tool crib, (c) the tool select filter option, (d) adding a 0.75in. hog nose cutter (ID: 170), (e) the hog nose cutter listed, and (f) the 0.75in. hog nose cutter selected

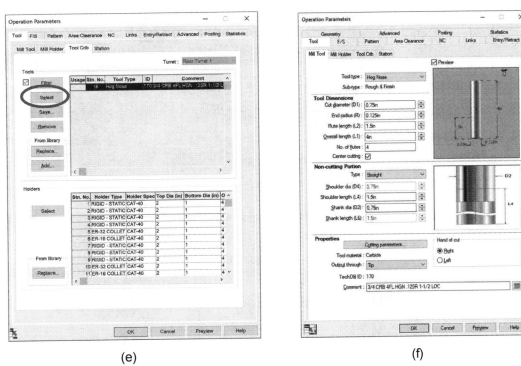

(e)　　　　　　　　　　　　　　　(f)

Figure 6.23 Replacing the tool with a 0.75in. hog nose (cont'd)

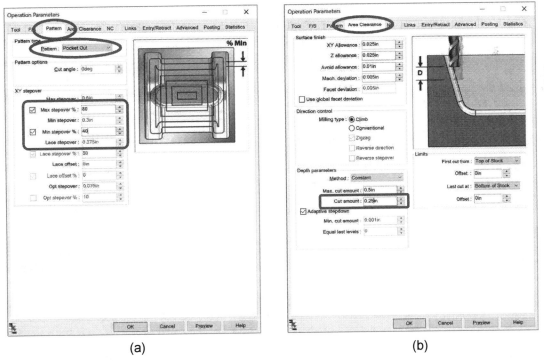

(a)　　　　　　　　　　　　　　　(b)

Figure 6.24 Modifying machining parameters, (a) modifying parameters and options under the *Pattern* tab, (b) changing *Cut amount* under the *Area Clearance* tab, and (c) showing machining time in the *Statistics* tab

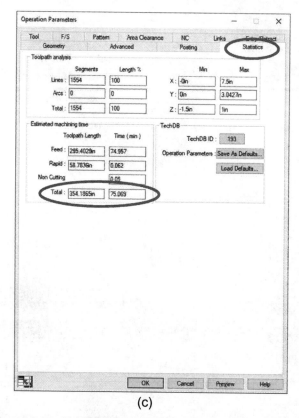

(c)

Figure 6.24 Modifying machining parameters (cont'd)

Although the machining time is reduced from 272 to 75.0 minutes, is the operation acceptable and is the stock-in-progress ready for the next operation, *Z Level*? Since a relatively large amount of material remained on the surface uncut due to a larger depth of cut (0.25 in.), we will have to use a smaller depth of cut (cut amount) for the follow-on *Z Level* operation.

6.6   Reviewing the Machined Part Quality

Now let us take a closer look at the machined part after the *Area Clearance1* operation. We will learn to use a few more command buttons in the *Simulate Toolpath* toolbox, in particular the *Section view* button to create section views.

Right click *Area Clearance1* and select *Simulate Toolpath*. The *Toolpath Simulation* toolbox appears; see Figure 4.30 of Lesson 4. Click the *Run* button ▶ to simulate the toolpath. The material removal simulation of the *Area Clearance1* operation will appear in the graphics area at the end, similar to that of Figure 6.25(a).

Click the *Stock Display* button 🖽 and choose *Wireframe Display* to display the stock in wireframe—see Figure 6.25(b)—which shows a better view in terms of the geometric shape of the machined surface in progress.

You may click the *Tool Display* button 🔨 and choose *No Display* to turn off the tool display, and click the *Target Part Display* button 🖽 and choose *Translucent Display*—see Figure 6.25(c)—showing the target part overlapping with the stock-in-progress; see Figure 6.25(d). Boundary edges of the target part shown in light green color clearly differentiate the significant amount of material remaining uncut after the *Area Clearance1* operation.

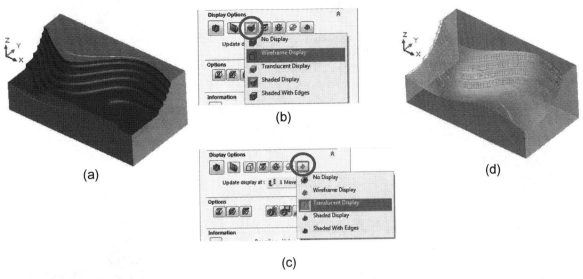

(a)

(b)

(c)

(d)

Figure 6.25 Options of the *Simulate Toolpath* tool box, (a) material removal simulation, (b) options to display stock in wireframe, (c) options to display target part in translucent, and (d) tool display turned off and target part shown in translucent display

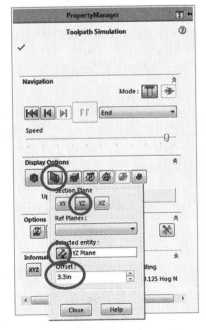

Figure 6.26 Defining a section view using the *Toolpath Simulation* toolbox

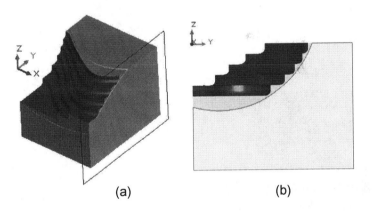

(a)

(b)

Figure 6.27 Section views of YZ plane, (a) current work view, and (b) right end view

You may create a section view to review the geometry of the machined part at specific sections normal to a direction of a selected plane. Click the *Section view* button ▨ from the *Toolpath Simulation* toolbox (Figure 6.26), choose *YZ* for *Plane*, click the *Reverse direction* button ▨, and increase the *Offset* to *3.3in*. (see Figure 6.26). A section view appears like that of Figure 6.27(a), which shows the difference between the target and the machined parts at the section.

You may rotate the view to see the section, for example, viewing it from the right end; see Figure 6.27(b). It is apparent that there is a significant amount of material uncut. Next, we modify the *Z Level* operation to remove the uncut material with a smaller tool and reduced stepover and depth of cut.

6.7   Reviewing and Modifying the Z Level Operation

As indicated in Figure 6.22(b), the machined surface after both *Area Clearance* and *Z Level* operations are not desirable and can be improved.

We right click *Z Level1* node under SOLIDWORKS CAM operation tree 🖳, and select *Edit Definition*.

In the *Operation Parameters* dialog box (see Figure 6.28), choose *Z Level* tab, and enter a smaller cut amount, for example, *0.025in*.

Regenerate toolpath for the *Z Level* operation, as seen in see Figure 6.29(a).

The toolpath seems to be fine except that the toolpath near the front edge of the freeform surface—circled in Figure 6.29(a)—may leave a noticeable amount of material uncut.

A similar observation is noted on the front right of the freeform surface. Note that the cutting time estimated for the revised *Z Level* operation is 15.7 minutes increased from 7.5 minutes before the change.

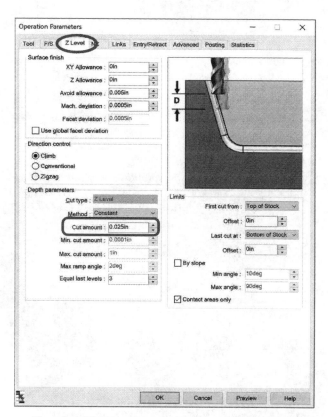

Figure 6.28 The *Z Level* tab of the *Operation Parameters* dialog box

Now, we simulate the toolpaths for the two operations combined: *Area Clearance1* and *Z Level1*.

Click the *Simulate Toolpath* button above the graphics area. Click the *Run* button ▶ to simulate the operations. The material removal simulation of the combined two operations appears in the graphics area, similar to that of Figure 6.29(b) and Figure 6.29(c) (section view). By visual inspection, the quality of the machined surface is in general very good, except that a noticeable amount of material uncut near the front edge—see the area circled in Figure 6.29(b)—is apparent, as seen before. Same is true around the area near the right side of the freeform surface of a less curvature, but less severe. The surface quality is much better comparing Figure 6.29(b) with Figure 6.22(b).

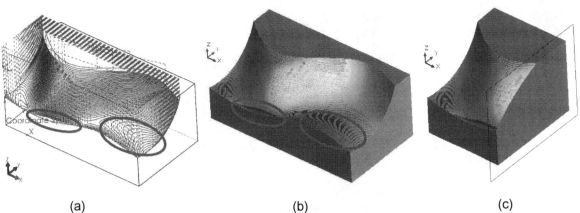

(a)   (b)   (c)

Figure 6.29 Toolpath and material removal simulation (cut amount: 0.025in.), (a) toolpath of *Z Level1*, (b) material removal simulation of the combined two operations, and (c) section view at offset 3.3in. in YZ plane

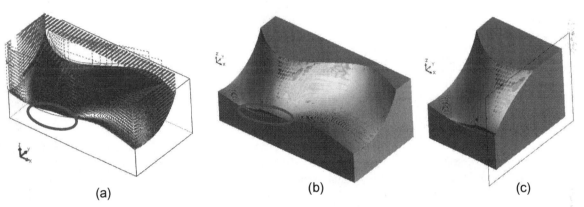

(a)   (b)   (c)

Figure 6.30 Toolpath and material removal simulation (cut amount: 0.01in.), (a) toolpath of *Z Level1*, (b) material removal simulation of the combined two operations, and (c) section view at offset 3.3in. in YZ plane

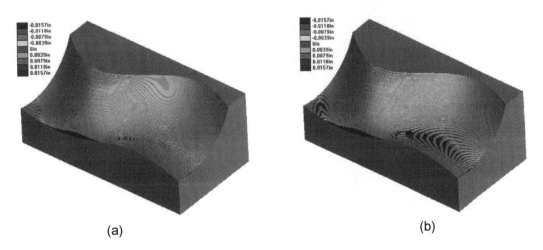

(a)   (b)

Figure 6.31 The material removal simulation of the *Show Difference* option showing differences between the machined part and the design part, (a) cut amount: 0.01in., and (b) cut amount: 0.025in.

How do we adjust the toolpath of the *Z Level1* operation to further improve the accuracy of the machined part? There is not much to change, except for further reducing the depth of cut.

Under SOLIDWORKS CAM operation tree 🄲 , right click *Z Level1* node and select *Edit Definition*. In the *Operation Parameters* dialog box (see Figure 6.28), choose *Z Level* tab, and enter a smaller cut amount, for example, *0.01in*. Click *OK* to regenerate the toolpath for the *Z Level1* operation. The toolpath and material removal simulation are shown in Figure 6.30. By visual inspection, the quality of the machined surface is improved, except for the material uncut near the front edge of the freeform surface. The flatter area on the right is much improved. The surface quality is better in general comparing Figure 6.30(b) with Figure 6.29(b).

The improvement is assured by choosing the *Show difference* button 🄲 under *Display Options*—again, circled in Figure 6.22(a)—to show a visual comparison of the machined part and the design part like that of Figure 6.31(a). Comparing Figure 6.31(a) with that of 0.025in. cut amount shown in Figure 6.31(b), the improvement is apparent (except for the material uncut near the front edge of the freeform surface).

Note that the cutting time estimated for the revised *Z Level* operation is 40.9 minutes (0.01in. cut amount), significantly increased from 15.7 minutes (0.025in. cut amount) before the change.

Even with such a small cut amount, there is still an excessive material uncut near the front edge of the freeform surface. This is because the curvature of the convex area near the front edge of the freeform surface makes the surface too steep for the tool to reach, leading to the noticeable uncut material. This problem can be properly addressed using a multiaxis surface milling operation. We will learn to generate operations using 5-axis mill in Lesson 14, and we will revisit this example at the time.

In summary, two major limitations of SOLIDWORKS CAM 2023 in cutting a freeform surface have been identified in this lesson, including (a) no surface milling operation that supports finish cut is available, and (b) no multiaxis surface milling is available that leads to significant uncut area near the front edge of the freeform surface in this example.

We have completed this exercise. You may save your model for future reference. Please keep the model file since we will need it again in Lesson 14.

6.8   Exercises

Problem 6.1. Follow the same steps discussed in this lesson to machine a part shown in Figure 6.32 using a stock of 12in.×5.5in.×3.5in. The stock material is assumed Steel 1005. Report the tools and machining options selected for individual operations. Also report machining times of individual and combined operations. Would operations of 3-axis mill give you a satisfactory machined part? Do you notice uncut areas like those in, for example, Figure 6.31?

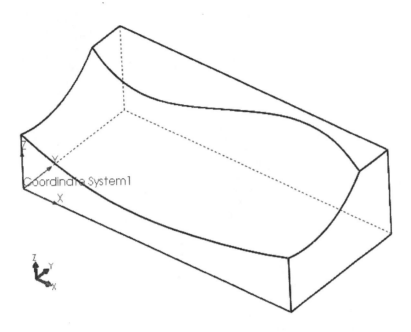

Figure 6.32 Solid model of Problem 6.1

[Notes]

Lesson 7: Multipart Machining

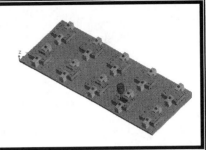

7.1   Overview of the Lesson

So far, we have discussed virtual machining for cutting a single part that is created as a part in SOLIDWORKS. In this lesson, we will move one step further, in which we focus on creating machining operations for a set of identical parts in an assembly created in SOLIDWORKS. Machining multiple parts in a single setup is a common practice at shop floor. Note that SOLIDWORKS CAM Professional version supports machining for an assembly. Standard version only supports part machining.

The individual part to be machined is identical to that in Lesson 5, which involves pocket milling and hole drilling using a 3-axis mill. The face milling operation is not included in this lesson to simplify the fixture design that holds the stock to a jig table. In SOLIDWORKS assembly, the part (more precisely, the stock) is assembled to a jig table by using two fixtures (one on each side). Each fixture consists of a clamp, a riser, and a threaded bolt. A total of ten parts arranged in two rows are to be machined, as shown in Figure 7.1.

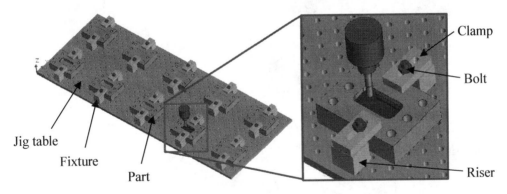

Figure 7.1 The material removal simulation of the multipart machining example

In this lesson, we learn the steps to create instances of the part in SOLIDWORKS CAM, define stocks for individual instances, extract machinable features, generate toolpath, and select components in the assembly (including the jig table and fixtures) for the tools to avoid. In addition, we will take a closer look at the G-code generated by SOLIDWORKS CAM for machining multiple parts in an assembly.

7.2   The Multipart Machining Example

The size of the bounding box of the part (filename: *2 point 5 axis features.SLDPRT*) is 8in.×6in.×2in. There is a center pocket and six holes, three on each side, as shown in Figure 7.2(a). As discussed in Lesson 5, these are 2.5 axis features that can be extracted automatically as machinable features by using the automatic feature recognition (AFR) capability. The stock size is identical to that of the bounding box.

As mentioned earlier, the face milling operation discussed in Lesson 5 is excluded in this lesson. The pocket milling operations, including two rough mills (Rough Mill1 and Rough Mill2) and a contour mill—toolpaths shown in Figure 7.2(b)—and hole drilling operations—center drill and drill shown in Figure 7.2(c)—are identical to those of Lesson 5.

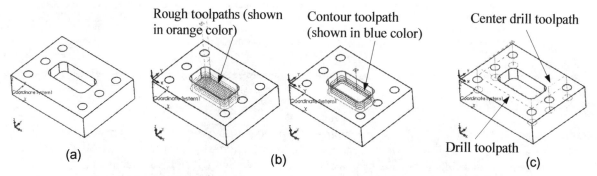

Figure 7.2 The *2 and 5 axis features* example: (a) part solid model, (b) pocket milling toolpaths: two rough and one contour, and (c) hole drilling toolpaths: center drill and drill

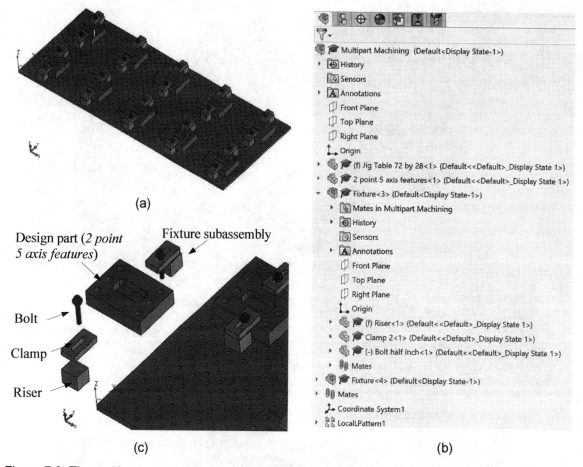

Figure 7.3 The multipart machining assembly example: (a) the entire assembly, (b) the feature tree, and (c) explode and zoom-in view

The stock is mounted on the jig table by two fixtures, one on each side. The jig table is 72in.×28in.×0.75in., with threaded holes of 0.5in. in diameter and 0.5in. in depth. The holes are 1.5in. apart every other row and column; that is, along the X and Y directions, respectively, of the coordinate system, *Coordinate System1*. This coordinate system is located at the front left corner of the jig table, as shown in Figure 7.3(a) and Figure 7.3(c).

The solid models *2 and 5 axis features.SLDPRT* are the only parts (as in ten instances) that will be machined. The fixtures are created as subassemblies; each consists of three parts, riser, clamp, and a bolt, as seen in Figure 7.3(c). In this lesson, we choose *Coordinate System1* for both fixture coordinate system and part setup origin. As a result, the G-code generated refers to this coordinate system. The design part (*2 and 5 axis features.SLDPRT*) with the two fixtures are patterned to create an additional nine instances in two rows as a linear pattern feature (*LocalPattern1* under the FeatureManager design tree 🌳). The unit system chosen is IPS (inch, pound, second). When you open the assembly model *Multipart Machining.SLDASM*, you should see 2 parts and 2 subassemblies, a coordinate system, and a pattern feature, listed in the feature tree like that of Figure 7.3(b).

You may open the example file with toolpath created (filename: *Multipart Machining with Toolpath.SLDASM*) to preview toolpath created for this example. When you open the file, you may expand the *Part Manager* node, expand *2 point 5 axis features.SLDPRT*, and then the *Instances* to see the ten instances of the part to be machined (*2 and 5 axis features<1>* to *<10>*), as shown in Figure 7.4.

Also, you should see the five operations listed under *Setup1* of the SOLIDWORKS CAM operation tree tab 🗂 (see Figure 7.4). You may click the *Simulate Toolpath* button 🔘 Simulate Toolpath above the graphics area to preview the machining simulation.

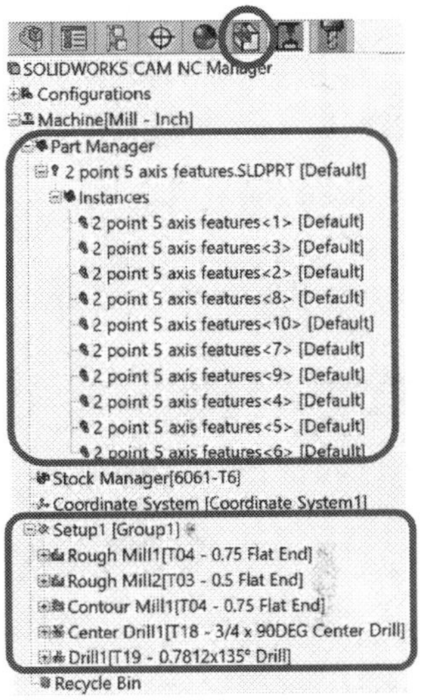

7.3 Using SOLIDWORKS CAM

Open SOLIDWORKS Assembly

Open the assembly model (filename: *Multipart Machining.SLDASM*) downloaded from the publisher's website. This assembly model, as shown in Figure 7.3, consists of four components (jig table, the part: *2 and 5 axis features*, and two fixtures), nine mates, a coordinate system, and a linear pattern. Again, as soon as you open the assembly model, you may want to check the unit system and make sure the IPS system is selected. You may also increase the decimals from the default 2 to 4 digits similar to that of previous lessons.

Select NC Machine

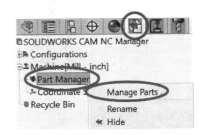

Figure 7.4 The instances and NC operations listed under the SOLIDWORKS CAM operation tree tab

Figure 7.5 Right clicking *Part Manager* and selecting *Manage Parts*

Click the SOLIDWORKS CAM feature tree tab 🗂 and right click *Mill-inch* to select *Edit Definition*. Similar to those of previous lessons, in the *Machine* dialog box, we select *Mill-inch* under *Machine* tab, choose *Tool Crib 2* under *Available tool cribs* of the *Tool Crib* tab, select *M3AXIS-TUTORIAL* under the *Post Processor* tab, and select *Coordinate System1* under *Fixture Coordinate System* of the *Setup* tab.

Manage Part

Since we are dealing with an assembly with multiple components, we have to identify which part or parts to cut. Under the SOLIDWORKS CAM feature tree tab 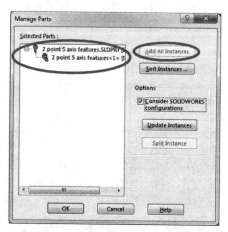, right click *Part Manager* and select *Manage Parts* (see Figure 7.5). The *Manage Parts* dialog box appears (see Figure 7.6, no entities selected initially).

Pick the part (*2 point 5 axis features*) in the graphics area or select *2 point 5 axis features* under the FeatureManager design tree tab . The part is now listed under *Selected Parts* in the *Manage Parts* dialog box (circled in Figure 7.6). Select *2 point 5 axis features* listed and click *Add All Instances* button to bring in all instances, and then click *OK* to accept the part definition.

Figure 7.6 The *Manage Parts* dialog box

Click the SOLIDWORKS CAM feature tree tab , expand the *Part Manager* node to see the part (*2 point 5 axis features.SLDPRT*) and its instances like that of Figure 7.7. Also, a *Stock Manager* node is added to the feature tree, as circled in Figure 7.7.

Create Stock for Instances

From the SOLIDWORKS CAM feature tree , right click *Stock Manager* and choose *Edit Definition*. In the *Stock Manager* dialog box (Figure 7.8), we use the default stock size (8in.×6in.×2in.) and choose material: *Aluminum 6061-T6*, and click the *Apply Current Definitions to All Parts* button (circled in Figure 7.8). Then click the checkmark ✓ to accept the stock definition.

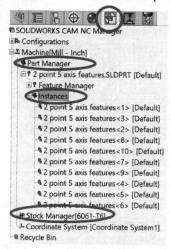

Figure 7.7 Entities listed under the SOLIDWORKS CAM feature tree tab

Figure 7.8 The *Stock Manager* dialog box

Extract Machinable Features

Click the *Extract Machinable Features* button above the graphics area. A *Setup* node is created with two machinable features extracted: *Rectangular Pocket* and *Hole Group* (with six holes), all listed in SOLIDWORKS CAM feature tree (Figure 7.9). Both machinable features are shown in magenta color.

In the graphics area (see Figure 7.10), a symbol with arrow pointing downward appears at the front left corner of the jig table coinciding with *Coordinate System1* (not shown in Figure 7.10). This symbol indicates that the tool axis is chosen correctly (pointing downward). You may need to move the mouse point over *Setup1* in the feature tree to show the part setup origin symbol in the graphics area (circled in Figure 7.10). This corner point also serves as the part setup origin, which locates G-code program zero.

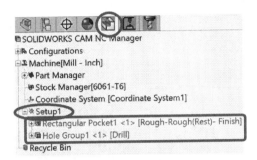

Figure 7.9 The two machinable features extracted

The default part setup origin

Figure 7.10 The default part setup origin coinciding with *Coordinate System1*

Generate Operation Plan and Toolpath

Click the *Generate Operation Plan* button above the graphics area. Five operations, Rough Mill1 (cutting the pocket), Rough Mill2 (cutting the corner fillets of the pocket), Contour Mill1 (cutting the boundary faces of the pocket), Center Drill1, and Drill1, are generated. They are listed in SOLIDWORKS CAM operation tree (Figure 7.11). Again they are shown in magenta color. Click the *Generate Toolpath* button above the graphics area to create the toolpath. The five operations are turned into black color after toolpaths are generated.

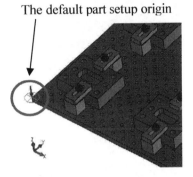

Figure 7.11 The five operations generated

Part Setup Origin

As pointed out earlier, if you click *Setup1* under SOLIDWORKS CAM operation tree , the part setup origin appears at the front left corner of the jig table, coinciding with the fixture coordinate system. Next, we review the options that define the setup origin for machining multiple parts in assembly.

Click the SOLIDWORKS CAM operation tree tab , and right click *Setup1* to choose *Edit Definition*. In the *Setup Parameters* dialog box, click the *Origin* tab (Figure 7.12). Choose *Setup origin* under *Output origin*, choose *Fixture coordinate system* under *Setup origin*, as circled in Figure 7.12, and leave *0* for *X:*, *Y:*, and *Z:* coordinates. This implies that the G-code will be output referring back to the SOLIDWORKS assembly coordinate system, which is *Coordinate System1*.

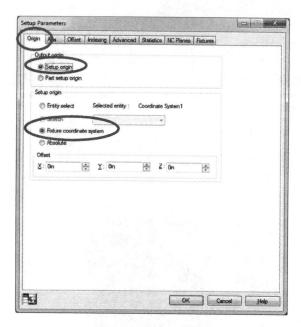

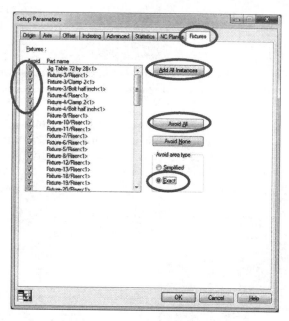

Figure 7.12 The *Origin* tab in the *Setup Parameters* dialog box

Figure 7.13 The *Fixtures* tab in the *Setup Parameters* dialog box

We review the G-code output by SOLIDWORKS CAM in Section 7.5, based on the selections. We will explore other options in the exercise problems at the end of the lesson.

Fixtures and Components to Avoid

We will select the jig table and the two fixtures to avoid in generating the toolpath.

In the *Setup Parameters* dialog box, click the *Fixtures* tab (Figure 7.13). Choose the components to avoid, including the jig table and riser, clamp, and bolt of the two fixture subassemblies, from the FeatureManager design tree tab 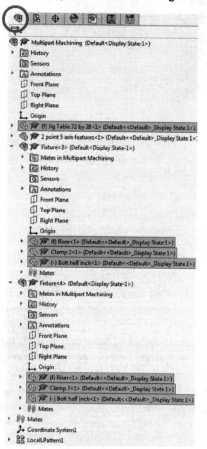 (see Figure 7.14).

In the *Setup Parameters* dialog box, choose all components listed (by clicking the small checkboxes in front, as circled in Figure 7.13), and click *Add All Instances*, *Avoid All*, and *Exact*, then click *OK*. Please make sure you select *Exact*. If not, you may see undesirable toolpath, for example, in Rough Mill2 operation. You may also need to choose one part at the time (by clicking the small box) and click *Add All Instances* to make sure all instances are included.

Figure 7.14 Selecting components to avoid

If you see a warning message: *The origin or machining direction or advanced parameters has changed, toolpaths need to be recalculated. Regenerate toolpaths now?* Click *Yes*. The toolpath will be regenerated like that shown in Figure 7.15.

7.4   The Sequence of Part Machining

You may click the *Simulate Toolpath* button to run the material removal simulation like that of Figure 7.1.

Note that the machining sequence follows a counterclockwise order looking down from above the jig table, starting from the part labeled 1 close to the front left corner, to part 10 close to the rear left corner, as shown in Figure 7.16.

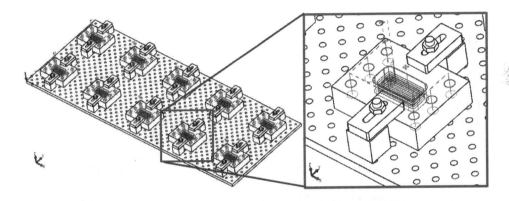

Figure 7.15   Toolpath of the multipart machining example

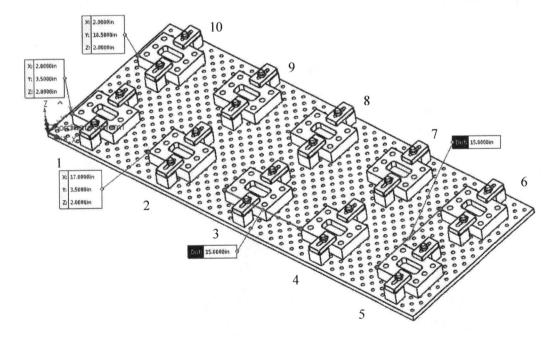

Figure 7.16   Locations of referencing corner points and distances between parts

The Rough Mill1 operation ran over all ten instances first, followed by Rough Mill2, Contour Mill, Center Drill, and Drill operations.

7.5   The G-code

Recall that in this example, we chose *Setup origin* for *Output origin* and *Fixture coordinate system* for *Setup origin* (see Figure 7.12). We expect that the G-code is output referring to the SOLIDWORKS assembly coordinate system, which is *Coordinate System1*, located at the front left corner of the top face of the jig table (as shown in Figure 7.10).

You may click the *Post Process* button ⌗ above the graphics area, and follow the same steps learned in the previous lessons (see, for example, Lesson 4) to convert the toolpaths into G-code.

Before reviewing the G-code, we acquire a basic understanding of the arrangement of the ten parts on the jig table. We also acquire a few referencing dimensions that help us verify the G-code.

As shown in Figure 7.16, the coordinates of the front left corner at the top face of stock 1 is (2, 3.5, 2), stock 2 is (17, 3.5, 2), and stock 10 is (2, 18.5, 2) referring back to *Coordinate System1*. The coordinates of the referencing points can be found by using the measure capability of SOLIDWORKS (by choosing from the pull-down menu *Tools > Evaluate > Measure*). Therefore, the distance between neighboring parts is 15in. along both X- and Y-directions, as shown in Figure 7.16.

Also, since the size of the block is 8in.×6in.×2in., the center points at the top face of the individual stocks are stock 1 (6, 6.5, 2), stock 2 (21, 6.5, 2), stock 3 (36, 6.5, 2), stock 4 (51, 6.5, 2), stock 5 (66, 6.5, 2), stock 6 (66, 21.5, 2), stock 7 (51, 21.5, 2), stock 8 (36, 21.5, 2), stock 9 (21, 21.5, 2), and stock 10 (6, 21.5, 2). All refer to *Coordinate System1*.

The X and Y coordinates of the hole centers, for example the three holes on the rear end of the pocket of stock 1, are respectively (3, 8.5), (3, 6.5), and (3, 4.5), from right to left [or Holes 1 to 3 shown in Figure 7.17(a)]. The other three holes on the front end of the pocket, from left to right, are respectively (9, 4.5), (9, 6.5), and (9, 8.5), or Holes 4 to 6 shown in Figure 7.17(b). All refer to *Coordinate System1*.

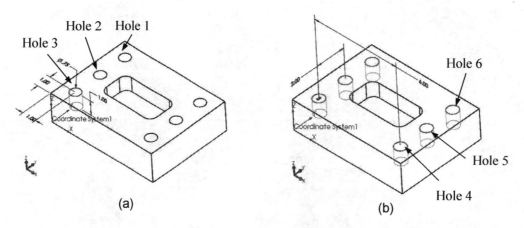

(a)   (b)

Figure 7.17 Dimensions of the hole centers of the design model: (a) the three holes on the rear end of the pocket, and (c) three holes on the front end of the pocket

```
O0001
N1 G20
N2 (3/4 EM CRB 2FL 1-1/2 LOC)
N3 G91 G28 X0 Y0 Z0
N4 T04 M06
N5 S3677 M03

N6 ( Rough Mill1 )
N7 G90 G54 G00 X4.985 Y6.515
N8 G43 Z2.1 H04 M08
N9 G01 Z1.625 F4.0448
N10 G17 Y6.485 F16.1793
N11 X7.015
N12 Y6.515
N13 X4.985
...

N107 ( Rough Mill1 )
N108 X19.985 Y6.515
N109 Z2.1
N110 G01 Z1.625 F4.0448
N111 Y6.485 F16.1793
N112 X22.015
N113 Y6.515
N114 X19.985
...

N208 ( Rough Mill1 )
N209 X34.985 Y6.515
N210 Z2.1
N211 G01 Z1.625 F4.0448
N212 Y6.485 F16.1793
N213 X37.015
N214 Y6.515
N215 X34.985
N216 Y6.815
N217 X34.685
N218 Y6.185
N219 X37.315
...

N1020 ( Rough Mill2 )
N1021 G90 G54 G00 X4.5 Y7.24
N1022 G43 Z2.1 H03 M08
N1023 G01 Z1.75 F6.8151
N1024 G03 X4.26 Y7. I0 J-.24 F27.2606
N1025 G01 Y6.98
N1026 G02 X4.52 Y7.24 I.26 J0
N1027 G01 X4.5
N1028 G00 Z2.1
...

N1135 ( Rough Mill2 )
N1136 X19.5
N1137 Z2.1
N1138 G01 Z1.75 F6.8151
N1139 G03 X19.26 Y7. I0 J-.24 F27.2606
...

N2174 ( Contour Mill1 )
N2175 G90 G54 G00 X6.3663 Y6.8273
N2176 G43 Z2.1 H04 M08
N2177 G01 Z1.625 F4.0448
N2178 G41 D24 X6.0905 Y7.103 F12.1345
N2179 G03 X6.0375 Y7.125 I-.053 J-.053
N2180 G01 X4.5 F16.1793
N2181 G03 X4.375 Y7. I0 J-.125
...
```

```
N2708 ( Center Drill1 )
N2709 G90 G54 G00 X3. Y8.5
N2710 G43 Z3. H17 M08
N2711 G82 G98 R2.75 Z1.6443 P00 F26.4528
N2712 Y6.5
N2713 Y4.5
N2714 X9.
N2715 Y6.5
N2716 Y8.5
N2717 G80 Z3.

N2718 ( Center Drill1 )
N2719 X18.
N2720 G82 G98 R2.75 Z1.6443 P00 F26.4528
N2721 Y6.5
N2722 Y4.5
N2723 X24.
N2724 Y6.5
N2725 Y8.5
N2726 G80 Z3.

N2727 ( Center Drill1 )
N2728 X33.
N2729 G82 G98 R2.75 Z1.6443 P00 F26.4528
N2730 Y6.5
N2731 Y4.5
N2732 X39.
N2733 Y6.5
N2734 Y8.5
N2735 G80 Z3.

N2736 ( Center Drill1 )
N2737 X48.
N2738 G82 G98 R2.75 Z1.6443 P00 F26.4528
N2739 Y6.5
N2740 Y4.5
N2741 X54.
N2742 Y6.5
N2743 Y8.5
N2744 G80 Z3.
...

2781 ( Center Drill1 )
N2782 X18.
N2783 G82 G98 R2.75 Z1.6443 P00 F26.4528
N2784 Y21.5
N2785 Y19.5
N2786 X24.
N2787 Y21.5
N2788 Y23.5
N2789 G80 Z3.

N2790 ( Center Drill1 )
N2791 X3.
N2792 G82 G98 R2.75 Z1.6443 P00 F26.4528
N2793 Y21.5
N2794 Y19.5
N2795 X9.
N2796 Y21.5
N2797 Y23.5
N2798 G80 Z3. M09
N2799 G91 G28 Z0
N2800 (25/32 SCREW MACH DRILL)
N2801 T18 M06
N2802 S4967 M03
```

```
N2803 ( Drill1 )
N2804 G90 G54 G00 X3. Y8.5
N2805 G43 Z3. H18 M08
N2806 G83 G98 R2.75 Z1. Q.1 F29.8067
N2807 Y6.5
N2808 Y4.5
N2809 X9.
N2810 Y6.5
N2811 Y8.5
N2812 G80 Z3.

N2813 ( Drill1 )
N2814 X18.
N2815 G83 G98 R2.75 Z1. Q.1 F29.8067
N2816 Y6.5
N2817 Y4.5
N2818 X24.
N2819 Y6.5
N2820 Y8.5
N2821 G80 Z3.

N2822 ( Drill1 )
N2823 X33.
N2824 G83 G98 R2.75 Z1. Q.1 F29.8067
N2825 Y6.5
N2826 Y4.5
N2827 X39.
N2828 Y6.5
N2829 Y8.5
N2830 G80 Z3.

N2831 ( Drill1 )
N2832 X48.
N2833 G83 G98 R2.75 Z1. Q.1 F29.8067
N2834 Y6.5
N2835 Y4.5
N2836 X54.
N2837 Y6.5
N2838 Y8.5
N2839 G80 Z3.
...

N2876 ( Drill1 )
N2877 X18.
N2878 G83 G98 R2.75 Z1. Q.1 F29.8067
N2879 Y21.5
N2880 Y19.5
N2881 X24.
N2882 Y21.5
N2883 Y23.5
N2884 G80 Z3.

N2885 ( Drill1 )
N2886 X3.
N2887 G83 G98 R2.75 Z1. Q.1 F29.8067
N2888 Y21.5
N2889 Y19.5
N2890 X9.
N2891 Y21.5
N2892 Y23.5
N2893 G80 Z3. M09
N2894 G91 G28 Z0
N2895 G28 X0 Y0
N2896 M30
```

Figure 7.18 The G-code, with no subroutine

Figure 7.18 shows (partial) contents of the G-code (O0001), consisting of over 2800 blocks, which is large. This is because every single operation is repeated ten times for the respective ten instances.

One easier way to verify the G-code is to review the cutter locations of the center drill and drill operations. The NC blocks of center drill and drill operations are listed in the second and third columns of Figure 7.18, respectively.

NC blocks N2708 to N2717 center drill the six holes of stock 1. Block N2709 moves the drill to Hole 1 (G00 X3. Y8.5), N2712 center drills Hole 2 (X3. Y6.5), and N2713 to N2716 center drill Holes 3 to 6, respectively.

NC blocks N2718 to N2726 center drill the six holes of stock 2. Block N2719 moves the drill to Hole 1 (X18. Y8.5), N2721 center drills Hole 2 (X18. Y6.5), and N2722 to N2725 center drill Holes 3 to 6, respectively.

You may review more NC blocks to find the cutter locations for the center drill operations on the remaining 8 stocks.

The NC blocks of drill operations are listed in the third column of Figure 7.18, respectively. The cutter locations of the drill operations are identical to those of center drills as they should be.

The G-code output by the SOLIDWORKS CAM post processor M3AXIS-TUTORIAL seems to be all good.

Although the G-code shown in Figure 7.18 is good, the code is lengthy. It is desirable to output G-code as subroutines for machining operations that cut part instances.

In SOLIDWORKS CAM, the option to output G-code with subroutines can be found under the *Posting* tab of the *Machine* dialog box (see Figure 7.19).

You may right click *Machine* in the SOLIDWORKS CAM operation tree 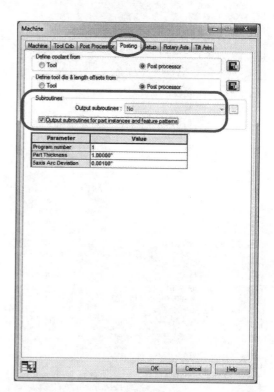 and choose *Edit Definition* to bring up the *Machine* dialog box. Choose the *Posting* tab. Click the checkbox in front of *Output subroutines for part instances and feature patterns* under *Subroutines* (circled in Figure 7.19) and click *OK*.

Click the *Post Process* button to convert the toolpath into G-code again. The G-code now consists of one main program: O0001, and five subroutines. They are O0002 (Rough Mill1), O0003 (Rough Mill2), O0004 (Contour Mill1), O0005 (Center Drill), and O0006 (Drill).

Figure 7.19 The *Posting* tab of the *Machine* dialog box

As shown in Figure 7.20, in addition to a few blocks at the beginning and the end, the main program is written roughly in five segments, Rough Mill1 (N6 to N50), Rough Mill2 (N51 to N95), Contour Mill1 (N96 to N140), Center Drill1 (N141 to N185), and Drill1 (N186 to N229).

In the first segment, Rough Mill1, a local coordinate system was assigned at the center point of the top face of individual stocks using the G52 NC word; for example, blocks N8 (stock 1, X6. Y6.5 Z2.), N12 (stock 2, X21. Y6.5 Z2.), to N44 (stock 10, X6. Y21.5 Z2.). Subprogram O0002 was called using M98 NC word right after the local coordinate system was set; for example, blocks N9 (stock 1), N13 (stock 2), to N45 (stock 10). G52 appears again in the main program—for example, N10, N14, N18, to N46, right after returning from subprogram to set the coordinate system back to (0,0,0).

The same can be found for the remaining operations: Rough Mill2, Contour Mill1, Center Drill1, and Drill1.

Next, we take a look at the subroutines, in particular Center Drill1 (O0005) and Drill1 (O0006), shown in the last column of Figure 7.21.

In Center Drill1 operation (O0005), the locations of the center of the six holes are specified in blocks N1 (Hole1: X-3. Y2.) and N4-N8 referring to the center points of Holes 2-6 at top face of the respective stock. Note that since the part setup origin was set at the top left corner of the jig table, the main program sets local coordinate systems to the center point of the top face of the respective stock by referring back to *Coordinate System1*. This is how the post process was written.

O0001	N51 (Rough Mill2)	N141 (Center Drill1)
N1 G20	N52 G90	N142 G90
N2 (3/4 EM CRB 2FL 1-1/2 LOC)	N53 G52 X6. Y6.5 Z2.	N143 G52 X6. Y6.5 Z2.
N3 G91 G28 X0 Y0 Z0	N54 M98 P0003	N144 M98 P0005
N4 T04 M06	N55 G52 X0 Y0 Z0	N145 G52 X0 Y0 Z0
N5 S3677 M03		
	...	...
N6 (Rough Mill1)		
N7 G90	N88 (Rough Mill2)	N178 (Center Drill1)
N8 G52 X6. Y6.5 Z2.	N89 G52 X6. Y21.5 Z2.	N179 G52 X6. Y21.5 Z2.
N9 M98 P0002	N90 M98 P0003	N180 M98 P0005
N10 G52 X0 Y0 Z0	N91 G52 X0 Y0 Z0	N181 G52 X0 Y0 Z0
	N92 G91 G28 Z0	N182 G91 G28 Z0
N11 (Rough Mill1)	N93 (3/4 EM CRB 2FL 1-1/2 LOC)	N183 (25/32 SCREW MACH DRILL)
N12 G52 X21. Y6.5 Z2.	N94 T04 M06	N184 T18 M06
N13 M98 P0002	N95 S3677 M03	N185 S4967 M03
N14 G52 X0 Y0 Z0		
	N96 (Contour Mill1)	N186 (Drill1)
N15 (Rough Mill1)	N97 G90	N187 G90
N16 G52 X36. Y6.5 Z2.	N98 G52 X6. Y6.5 Z2.	N188 G52 X6. Y6.5 Z2.
N17 M98 P0002	N99 M98 P0004	N189 M98 P0006
N18 G52 X0 Y0 Z0	N100 G52 X0 Y0 Z0	N190 G52 X0 Y0 Z0
...	...	...
N43 (Rough Mill1)	N133 (Contour Mill1)	N223 (Drill1)
44 G52 X6. Y21.5 Z2.	N134 G52 X6. Y21.5 Z2.	N223 (Drill1)
N45 M98 P0002	N135 M98 P0004	N224 G52 X6. Y21.5 Z2.
N46 G52 X0 Y0 Z0	N136 G52 X0 Y0 Z0	N225 M98 P0006
N47 G00 G91 G28 Z0	N137 G91 G28 Z0	N226 G52 X0 Y0 Z0
N48 (1/2 EM CRB 2FL 1 LOC)	N138 (3/4 X 90DEG CBT SPOT	N227 G91 G28 Z0
N49 T03 M06	DRILL)	N228 G28 X0 Y0
N50 S6195 M03	N139 T17 M06	N229 M30
	N140 S4991 M03	

Figure 7.20 The G-code, main program

Similarly, for Drill1 operations, the main program (third column in Figure 7.20) sets local coordinate systems (G52) and calls subprogram O0006 (M98). The contents of subprogram O0006 shown in the last column of Figure 7.21 are similar to those of subprogram O0005.

For the pocket milling operations, as shown in the first two columns of Figure 7.20, the main program sets local coordinate systems (G52) and calls subprogram O0002 for Rough Mill1, O0003 for Rough Mill2, and then O0004 for Contour Mill1 (M98). The partial contents of subprograms O0002, O0003, and O0004 are shown in the first three columns of Figure 7.21. The X- and Y-coordinates of the cutter locations shown in the subprograms O0002, O0003, and O0004 are referred to the respective local coordinate systems, which are again the center point at the top face of the respective stocks, set by using G52 in the main program.

We have verified that G-code is correctly generated. We have now completed this exercise. You may save your model for future references.

O0002	O0003	O0004	O0005
N1 G90 G54 G00 X-1.015 Y.015	N1 G90 G54 G00 X-1.5 Y.74	N1 G90 G54 G00 X.3663 Y.3273	N1 G90 G54 G00 X-3. Y2.
N2 G43 Z1. H04 M08	N2 G43 Z1. H03 M08	N2 G43 Z1. H04 M08	N2 G43 Z1. H17 M08
N3 Z.1	N3 Z.1	N3 Z.1	N3 G82 G98 R.75 Z-.3557 P00
N4 G01 Z-.375 F4.0448	N4 G01 Z-.25 F6.8151	N4 G01 Z-.375 F4.0448	F26.4528
N5 G17 Y-.015 F16.1793	N5 G17 G03 X-1.74 Y.5 I0 J-.24	N5 G41 D24 X.0905 Y.603 F12.1345	N4 Y0
N6 X1.015	F27.2606	N6 G17 G03 X.0375 Y.625 I-.053 J-.053	N5 Y-2.
N7 Y.015	N6 G01 Y.48	N7 G01 X-1.5 F16.1793	N6 X3.
N8 X-1.015	N7 G02 X-1.48 Y.74 I.26 J0	N8 G03 X-1.625 Y.5 I0 J-.125	N7 Y0
N9 Y.315	N8 G01 X-1.5	N9 G01 Y-.5	N8 Y2.
N10 X-1.315	N9 G00 Z.1	N10 G03 X-1.5 Y-.625 I.125 J0	N9 G80 Z1. M09
N11 Y-.315	N10 Z-.15	N11 G01 X1.5	N10 M99
N12 X1.315	N11 G01 Z-.5 F6.8151	N12 G03 X1.625 Y-.5 I0 J.125	
N13 Y.315	N12 G03 X-1.74 Y.5 I0 J-.24 F27.2606	N13 G01 Y.5	
N14 X-1.015	...	N14 G03 X1.5 Y.625 I-.125 J0	
N15 Y.615		...	O0006
N16 X-1.5	N108 Z-.65		N1 G90 G54 G00 X-3. Y2.
...	N109 G01 Z-1. F6.8151	N47 G01 Y.5	N2 G43 Z1. H18 M08
	N110 G02 X1.74 Y.48 I0 J-.26 F27.2606	N48 G03 X1.5 Y.625 I-.125 J0	N3 G83 G98 R.75 Z-1. Q.1 F29.8067
N96 G03 X1.615 Y-.5 I0 J.115	N111 G01 Y.5	N49 G01 X-.0375	N4 Y0
N97 G01 Y.5	N112 G03 X1.5 Y.74 I-.24 J0	N50 G03 X-.0905 Y.603 I0 J-.075	N5 Y-2.
N98 G03 X1.5 Y.615 I-.115 J0	N113 G01 X1.48	N51 G40 G01 X-.3663 Y.3273	N6 X3.
N99 G01 X-1.015	N114 G00 Z.1	N52 G00 Z.1	N7 Y0
N100 G00 Z.1	N115 Z1. M09	N53 Z1. M09	N8 Y2.
N101 Z1. M09	N116 M99	N54 M99	N9 G80 Z1. M09
N102 M99			N10 M99

Figure 7.21 The G-code of subprograms, O0002, O0003, O0004, O0005, and O0006

7.6   Exercises

Problem 7.1.   Create an assembly like that of Figure 7.22 using the parts (*Jig table*, *Problem 7.1 Part*, *Clamp*, and *Short Bolt*) and subassembly (*Fixture*) in the Exercises folder of Lesson 7 downloaded from the publisher's website. Note that *Problem 7.1 Part* is identical to that of Problem 6.1. Create a total of eight instances as a linear pattern feature.

(a)   Generate a machining simulation for the eight instances using the machining operations created in Problem 6.1.

(b)   Generate G-code with selections like those discussed in this lesson. Verify that the codes are generated correctly by reviewing the contents of the codes, similar to those of Section 7.5.

Hint: You may perform the following to manually create a machinable feature in assembly, (a) create instances by right clicking *Part Manager* and choosing *Manage Parts*, (b) insert a mill part setup by expanding *Problem 7.1 Part* and right clicking *Feature Manager*, and (c) insert a new multi surface feature by expanding *Feature Manager* and right clicking *Mill Part Setup*.

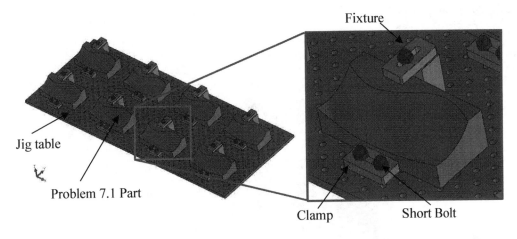

Fixture

Jig table

Problem 7.1 Part

Clamp

Short Bolt

Figure 7.22 The assembly model of Problem 7.1

Problem 7.2. Now we get back to the example of this lesson to explore other options in selecting part setup origin. Bring out the *Setup Parameters* dialog box and select *Part setup origin* for *Output origin* (see Figure 7.23). Note that the part setup origin symbols appear at the front left corner at the top face of individual stocks, as shown in Figure 7.24.

Output G-code. Open the G-code file and identify the differences that this selection makes to the G-code ouput in Section 7.5. Verify if the G-code are output correctly.

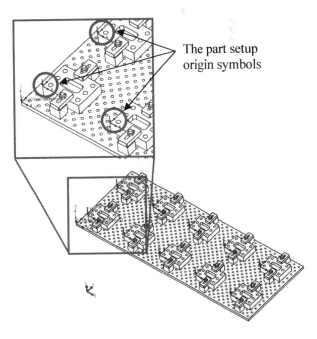

The part setup origin symbols

Figure 7.23 The *Origin* tab in the *Setup Parameters* dialog box

Figure 7.24 The part setup origin symbols

Problem 7.3. Continue from Problem 7.2. In this case, we choose *Setup origin* for *Output origin* (see Figure 7.25), click *Entity select* under *Setup origin*, and pick *Coordinate System1* shown in Figure 7.26. Then, we output G-code. Open the G-code and identify the differences that such options make to the G-code output in Section 7.5. Verify if the G-code are output correctly.

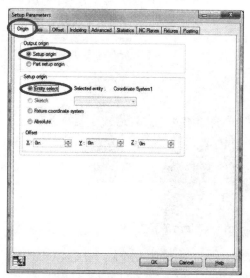

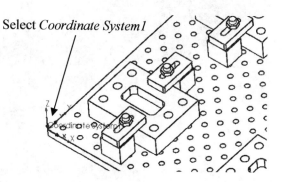

Figure 7.26 Select a corner point at part setup origin

Figure 7.25 The *Origin* tab in the *Setup Parameters* dialog box

Lesson 8: Multiplane Machining

8.1   Overview of the Lesson

In this lesson, we learn to create machining operations that cut parts with machinable features on multiple planes. In particular, we assume the stocks to be mounted on a tombstone that rotates with desired angles by a rotary table rotating along the longitudinal direction; that is, the 4th axis or A-axis in this case.

The design part is similar to that of Lessons 5 and 7, except that there are holes on the side faces, in addition to the pocket and six holes on the top face. Similar to Lesson 7, the face milling discussed in Lesson 5 is not included to simplify the fixture design for stock setup. The part is assembled to a tombstone by using two identical fixtures (one on each end). Each fixture consists of a clamp and two bolts. A total of four parts mounted on the respective four faces of the tombstone are to be machined, as shown in Figure 8.1. The tombstone is mounted on a rotary table using four bolts, and the rotary table is assembled to a rotary unit, which is mounted on a jig table like that of Lesson 7 (not shown in Figure 8.1). We assume that a 3-axis mill with a rotary table is employed for this lesson. The rotary table rotates ±90° or ±180° along the longitudinal direction for the tools to cut features on faces that are perpendicular to the tool axis.

We will learn the steps to define the 4th axis, in addition to creating instances of part for generating machining operations that are similar to that of Lesson 7, including defining stock for individual instances, extracting machinable features, generating toolpath, and selecting components in the assembly for the tools to avoid. We will take a closer look at the G-code generated by SOLIDWORKS CAM to verify that the code is generated correctly in support of the machining operations.

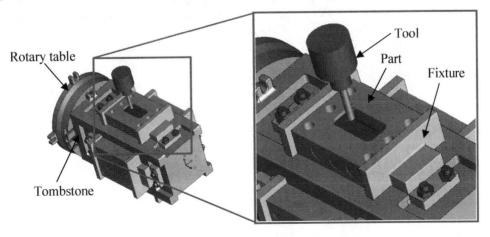

Figure 8.1 Material removal simulation of the multiplane machining example

8.2   The Multiplane Machining Example

A design part (filename: *2 point 5 axis features with side holes.SLDPRT*) similar to that of Lessons 5 and 7 with a bounding box of 8in.×6in.×2in. is employed in this lesson. In addition to the pocket and holes on its top face—see Figure 8.2(a)—there are three holes on its front side face, as shown in Figure 8.2(a), and another three holes on its rear side face; see Figure 8.2(b). All these holes on the side faces are of the same size, and they are extracted as machinable features by using the automatic feature recognition (AFR) capability.

The assembly model (*Multiplane Machining.SLDASM*) consists of eight parts, one subassembly (called *clamped part*), a circular pattern feature that includes the instances of the clamped part subassembly, and a point and a coordinate system, as listed in the FeatureManager design tree ![icon] shown in Figure 8.3(a).

The jig table of Lesson 7 is employed again for this lesson. A rotary unit is mounted on the jig table; see Figure 8.3(b). And a rotary table is assembled to the rotary unit by using four 0.5in. bolts, and a tombstone on which parts are mounted is assembled to the rotary table. The clamped part subassembly consisting of the design part (*2 and 5 axis features with side holes*), two fixtures, and two bolts are mounted on the respective four faces of the tombstone; see Figure 8.3(c).

A coordinate system, *Coordinate System1*, is defined at the center of the front end face of the tombstone with X-axis pointing along the longitudinal direction of the jig table; see Figure 8.3(b). Note that *Coordinate System1* will be chosen as the fixture coordinate system for the machining operations in this lesson. The rotary table provides rotation motion along the X-axis of the fixture coordinate system.

You may open the example file with toolpaths created (filename: *Multiplane Machining with Toolpath.SLDASM*) to preview the setups and toolpaths of this example.

You may click the SOLIDWORKS CAM feature tree tab ![icon] to preview the setups. There are four setups corresponding to the respective four tool axes defined in this example. As shown in Figure 8.4(a), the tool axis (symbol: ![icon]) of *Setup1* points in the –Z direction with four machinable features: *Rectangular Pocket1* and *Hole Group1* on the top face of the part mounted on the top face of the tombstone, *Hole Group2* on the side face of the part mounted on the front side face of the tombstone, and *Hole Group3* on the part mounted on the rear side face. As another example, the tool axis of *Setup3* points in the +Y direction—see Figure 8.4(b)—with another four machinable features, *Rectangular Pocket1* and *Hole Group1* on the part mounted on the front side face of the tombstone, *Hole Group2* on the part mounted on the bottom face, and *Hole Group3* on the part of the top face of the tombstone.

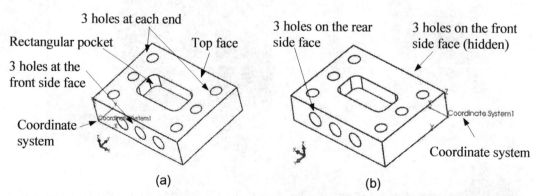

Figure 8.2 Machinable features of the design part, *2 and 5 axis features with side holes*, (a) features at the top and front side faces, and (b) holes on the rear side face (a rotated view)

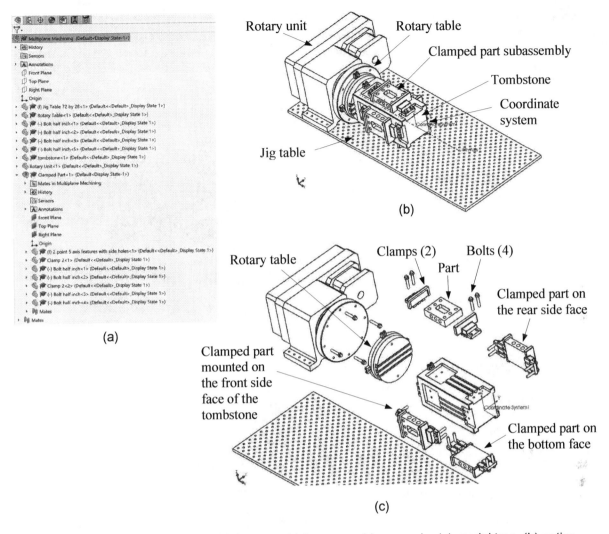

Figure 8.3 The solid model of multiplane machining assembly example: (a) model tree, (b) entire assembly, and (c) explode view

Tool axes of *Setup2* and *Setup4* point in $+Z$ and $-Y$ directions, respectively. Machinable features associated with these two setups are pockets and holes sketched on the part faces that are normal to the respective tool axes, similar to those of *Setup1* and *Setup3* shown in Figure 8.4.

Note that you might see the four setups in a different order as in the example file when you go through the exercise of this lesson.

The stock size is identical to that of the bounding box of the part. The pocket milling operations, (including two rough milling and a contour milling) and hole drilling operations (center drill and drill) are identical to those of Lesson 7 (for example, the first five operations of *Setup1* shown in Figure 8.5). Similarly, the drilling operations for holes of parts mounted on the side faces of the tombstone consist of a center drill and a drill operation (see Figure 8.5). The toolpaths of *Setup1* are shown in Figure 8.5 for illustration. Similar toolpaths can be found for the other three setups. Note that the part setup origin coincides with that of the origin of *Coordinate System1*, as pointed out in Figure 8.5.

While opening the assembly file, *Multiplane Machining with Toolpath.SLDASM*, you may click the SOLIDWORKS CAM operation tree tab ![icon], expand the *Part Manager* node, expand the part (*2 point 5 axis features with side holes.SLDPRT*), and then expand the *Instances* to see the four instances of the part to be cut (*Clamped part-1* to *Clamped part-4* shown in Figure 8.6). Also, you should see the nine operations listed under *Setup1*, as shown in Figure 8.6. Similar operations can be seen for the other three setups simply by expanding the respective entities in the operation tree.

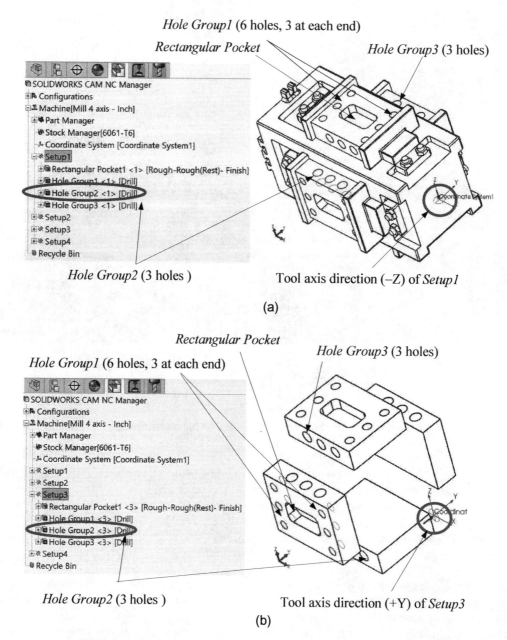

Figure 8.4 The setups generated for the multiplane machining example: (a) *Setup1* and associated machinable features, and (b) *Setup3* and associated machinable features (tombstone and fixtures not shown)

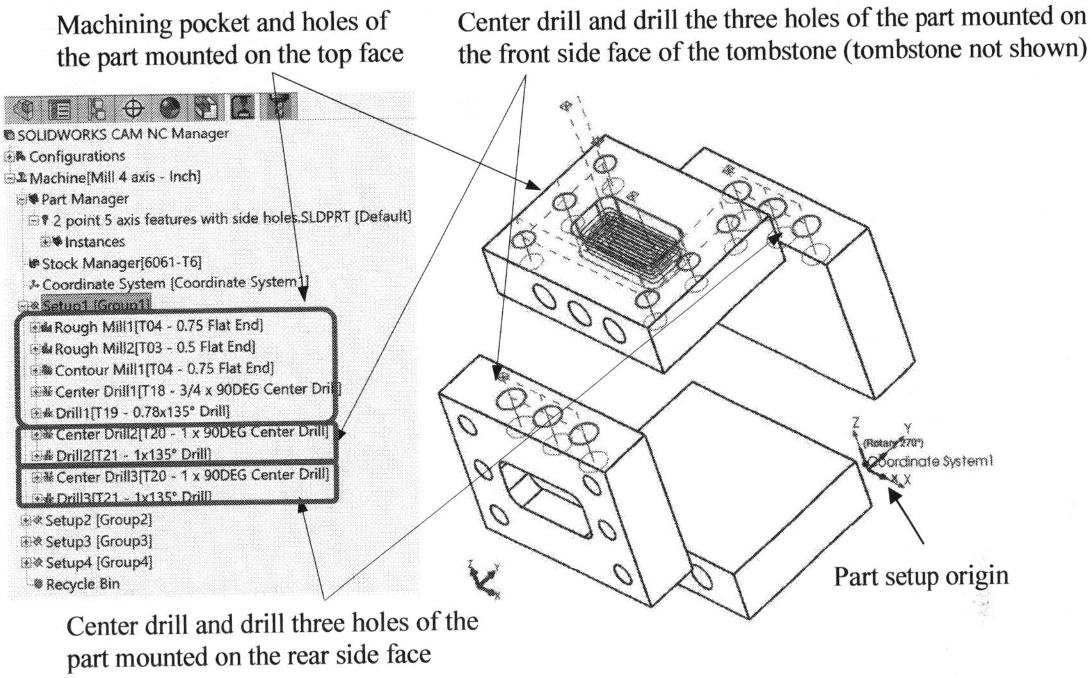

Machining pocket and holes of
the part mounted on the top face

Center drill and drill the three holes of the part mounted on
the front side face of the tombstone (tombstone not shown)

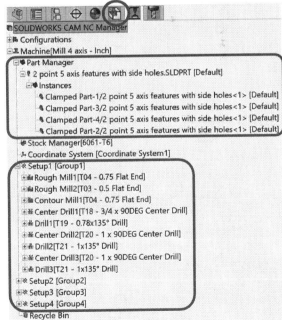

Part setup origin

Center drill and drill three holes of the
part mounted on the rear side face

Figure 8.5 Toolpaths of the nine operations of *Setup1* (tombstone and fixtures not shown)

You may click the *Simulate Toolpath* button above the graphics area to preview the machining operations.

8.3   Using SOLIDWORKS CAM

Open SOLIDWORKS Assembly

Open the assembly model (filename: *Multiplane Machining.SLDASM*) downloaded from the publisher's website. This assembly model appears in the graphics area similar to that of Figure 8.3(b). Again, as soon as you open the model, you may want to check the unit system chosen and make sure the IPS system is selected. You may also increase the decimals from the default 2 to 4 digits similar to that of previous lessons.

Select NC Machine

Click the SOLIDWORKS CAM feature tree tab 🗔 and right click *Mill-inch* to select *Edit Definition*. In the *Machine* dialog box, we choose *Mill 4 axis-inch* under the *Machine* tab; see Figure 8.7(a). Click *Select*.

Figure 8.6 The instances and NC operations
listed under the SOLIDWORKS CAM
operation tree tab

Choose *Tool Crib 2* under *Available tool cribs* of the *Tool Crib* tab, select *M4AXIS-TUTORIAL* under the *Post Processor* tab—Figure 8.7(b), and select *Coordinate System1* under the *Fixture Coordinate System* of the *Setup* tab.

Define the Rotary Axis

In the *Machine* dialog box, click the *Rotary Axis* tab, choose *X axis* as the *Rotary axis* and *XZ plane* as the *0 degree position*—circled in Figure 8.8(a). Then, click *OK*. In the graphics area, a circular arc with a counterclockwise arrow appears at the origin of *Coordinate System1*, as shown in Figure 8.8(b), indicating the rotation direction of the rotary axis. Click *OK* to accept the machine definition.

Manage Part

Similar to Lesson 7, we are dealing with an assembly of multiple design parts. Therefore, we have to assign which part or parts to cut. Under the SOLIDWORKS CAM feature tree tab 📑, right click *Part Manager* and select *Manage Parts* (Figure 8.9). The *Manage Parts* dialog box appears (Figure 8.10).

Pick the design part (*2 point 5 axis features with side holes*) in the graphics area or expand *Clamped Part*, and select *2 point 5 axis features with side holes* under the FeatureManager design tree tab 📑. The part is now listed under *Selected Parts* in the *Manage Parts* dialog box (Figure 8.10). Click *Add All Instances* button to bring in all instances. Click *OK* to accept the part definition.

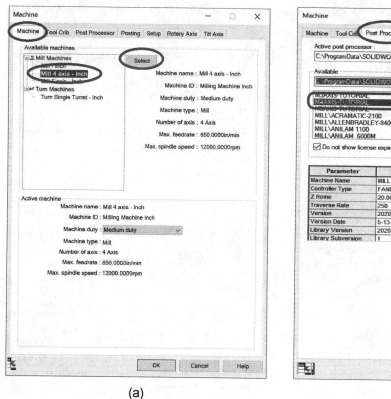

(a)

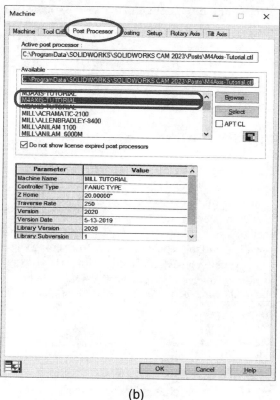

(b)

Figure 8.7 The *Machine* dialog box, (a) selecting 4 axis mill (*Mill 4 axis - inch*), and (b) selecting the 4 axis post-processor (*M4AXIS-TUTORIAL*)

Click the SOLIDWORKS CAM feature tree tab 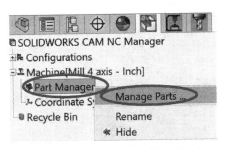, expand the *Part Manager* node to see the part (*2 point 5 axis features with side holes.SLDPRT*) and its instances to make sure that all four parts are included, as shown in Figure 8.11. Also, a *Stock Manager* node has been added.

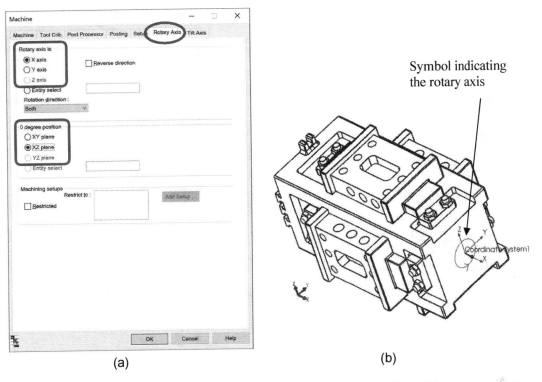

(a)   (b)

Figure 8.8 Defining the rotary axis, (a) the *Rotary Axis* tab of the *Machine* dialog box, and (b) the symbol indicating the rotary axis

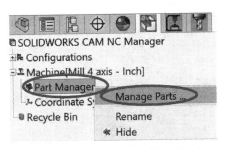

Figure 8.9 Right clicking *Part Manager* and select *Manage Parts*

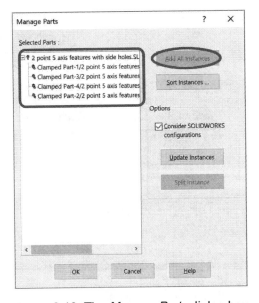

Figure 8.10 The *Manage Parts* dialog box

Create Stock for Part Instances

From SOLIDWORKS CAM feature tree 🔧 , right click *Stock Manager* and choose *Edit Definition*. In the *Stock Manager* dialog box (Figure 8.12), we leave the default stock size (8in.×6in.×2in.) and choose *6061-T6* for stock material, and click the *Apply Current Definitions to All Parts* button 🔩 (circled in Figure 8.12). Then click the checkmark ✔ to accept the definition of stock.

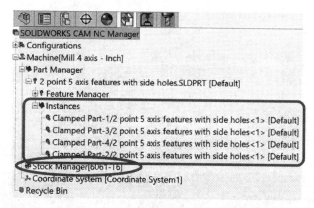

Figure 8.11 All four instances of the target part listed
under SOLIDWORKS CAM feature tree

Extract Machinable Features

Click the *Extract Machinable Features* button [Extract Machinable Features] above the graphics area. Four setup entities (*Setup1-4*) are created with four machinable features extracted per setup: *Rectangular Pocket1*, *Hole Group1* (with six holes), *Hole Group2* (with three holes), and *Hole Group3* (with another three holes), listed in SOLIDWORKS CAM feature tree 🔧 (see Figure 8.14). All machinable features are shown in magenta color.

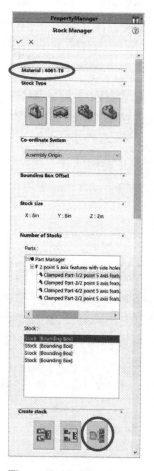

Figure 8.12 The *Stock Manager* dialog box

Click a setup to locate its part setup origin and tool axis in the graphics area. For example, click *Setup2* to see a symbol 🔾 with arrow pointing upward (coinciding with the +Z-axis) appearing at the center of the front end face of the tombstone (coinciding with *Coordinate System1*, as shown in Figure 8.13). All four setups shared the same part setup origin but with different tool axes. Tool axes of the four setups are pointing respectively in –Z, +Z, +Y, and –Y directions. Note that you may see setups with a different order. The order does not affect the end results, as long as you have all four setups, including ±Z and ±Y, in place.

Generate Operation Plan and Toolpath

Click the *Generate Operation Plan* button [Generate Operation Plan] above the graphics area. Nine operations, *Rough Mill1*, *Rough Mill2*, *Contour Mill1*, *Center Drill1*, *Drill1*, *Center Drill2*, *Drill2*, *Center Drill3*, and *Drill3*, are generated for each setup. There is a total of 36 operations, nine per setup. They are listed in SOLIDWORKS CAM operation tree 🗚 (see Figure 8.15, showing operations of *Setup1* and *Setup4*).

Again they are shown in magenta color. Click the *Generate Toolpath* button 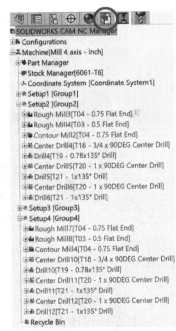 above the graphics area to create the toolpath. The operations are turned into black color after toolpaths are generated.

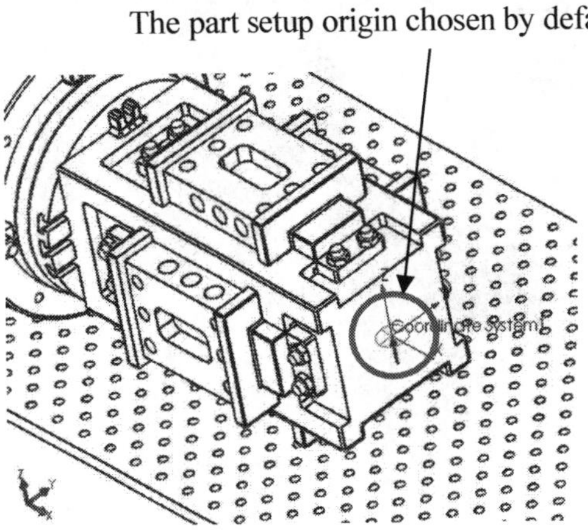

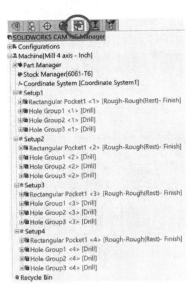

The part setup origin chosen by default

Figure 8.13 The default part setup origin coinciding with *Coordinate System1*

Figure 8.14 The machinable features extracted

Fixtures and Components to Avoid

Similar to Lesson 7, we select components to avoid in generating the toolpath. In this lesson, we select the rotary table, tombstone, fixtures, and bolts to avoid. We will not select the rotary unit and the jig table since the chance that the tools collide with them is minimum for a 3-axis mill with a rotary table.

On the other hand, only the components selected as fixtures are included in the material removal simulation. Excluding the jig table and the rotary unit makes the simulation more visually logical and realistic.

Click the SOLIDWORKS CAM operation tree tab, and right click *Setup1* to choose *Edit Definition*. In the *Setup Parameters* dialog box, click the *Features* tab (Figure 8.16). Choose the components to be part of the fixtures, including the rotary table, tombstone, both clamps, and bolts, either from the graphics area or from the FeatureManager design tree tab (see Figure 8.17) by clicking them.

Figure 8.15 The nine operations generated per setup

Also, click both clamps and bolts from the remaining three sets, including those of the circular pattern feature of the solid model. You may expand the pattern feature (*LocalCirPattern1* under FeatureManager design tree) to select the clamps and bolts. After selecting all these components, we click the *Avoid All* button to avoid them in the toolpath generation, then we click *OK*.

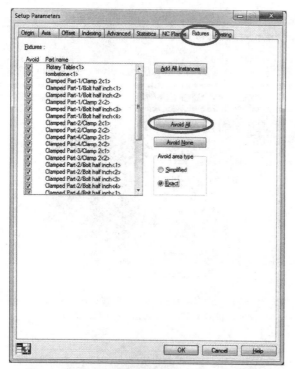

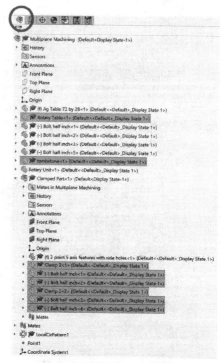

Figure 8.16 The *Features* tab in the *Setup Parameters* dialog box

Figure 8.17 Selecting the components to avoid

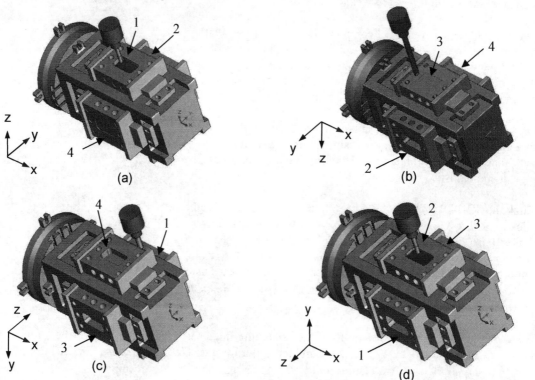

Figure 8.18 Material removal simulation (stock rotates), (a) *Setup1*, (b) *Setup2*, (c) *Setup3*, and (d) *Setup4*

Click *Yes* to the question in the warning box: *The origin or machining direction or advanced parameters has changed, toolpaths need to be recalculated. Regenerate toolpaths now?* The toolpaths will be regenerated, for example, like that shown in Figure 8.5 for *Setup1*.

Note that sometimes unintended toolpaths may be generated by SOLIDWORKS CAM, for example, cutters reaching areas outside the pocket, as occasionally observed in Rough Mill2 and Contour Mill1 of *Setup1*. All you have to do is to regenerate toolpath by right clicking these NC operations and selecting *Generate Toolpath*.

8.4 The Sequence of Part Machining

You may click the *Simulate Toolpath* button to run the material removal simulation like that of Figure 8.1.

Note that the machining sequence follows that of the four setups, from *Setup1* to *Setup4*. The rotary table rotates along the X-axis of *Coordinate System1*. The nine machining operations of *Setup1* cut machinable features on parts labelled 1, 2, and 4, as shown in Figure 8.18(a). The first three operations cut the pocket (2 rough and 1 contour mills) on part 1, followed by two operations (center drill and hole drilling) that drill the six holes on the top face of part 1. Then two operations drill the three holes (center drill and hole drilling) on part 4. The final two operations cut the three holes on part 2. The nine operations repeat three more times for the remaining respective three setups, as shown in Figure 8.18(b), (c), and (d) for *Setup2*, *Setup3*, and *Setup4*, respectively.

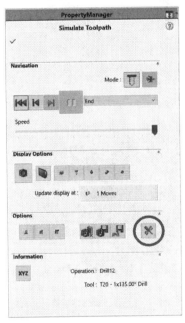

Figure 8.19 Click the *Options* button in the *Toolpath Simulation* toolbox

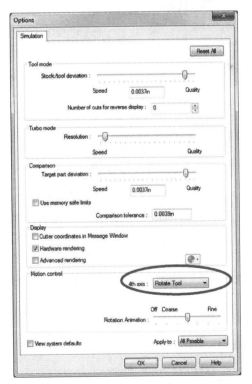

Figure 8.20 The *Features* tab in the *Setup Parameters* dialog box

In the default setting of the material removal simulation, stock rotates in the material removal simulation. The rotary table rotates a –180° angle from *Setup1* to *Setup2*, a 90° angle from *Setup2* to *Setup3*, and then a 180° angle from *Setup3* to *Setup4*.

You may change the default setting to rotate the tool in the material removal simulation (although this is less desirable from a practical perspective).

This can be done by clicking the *Options* button ⊠ in the *Toolpath Simulation* toolbox (Figure 8.19). In the *Options* dialog box (Figure 8.20), choose *Rotate Tool* for *4th axis* under *Motion control*, and then click *OK*.

Click *Yes* to the message: *Simulation must be restarted before new settings will take effect. Do you want to restart simulation?*

Click the *Run* button in the *Toolpath Simulation* toolbox. The material removal simulation will take place for all four setups, in which the tools rotate like those of Figure 8.21.

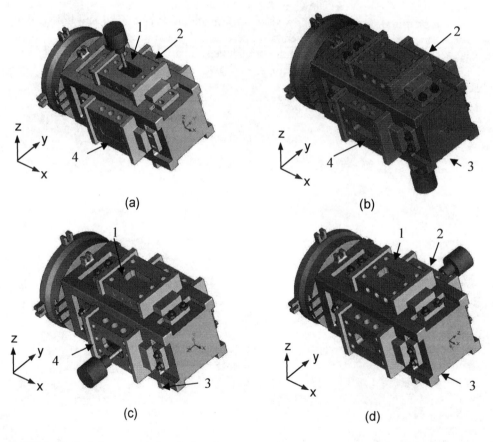

(a)   (b)

(c)   (d)

Figure 8.21 Material removal simulation (tool rotates), (a) *Setup1*, (b) *Setup2*, (c) *Setup3*, and (d) *Setup4*

8.5 Reviewing The G-code

In this example, we chose the part setup origin at the center point of the front end face of the tombstone, coinciding with that of the fixture coordinate system, *Coordinate System1*, as shown in Figure 8.4. We expect that the G-code generated by SOLIDWORKS CAM refers to the part setup origin at the coordinate system. In addition, recall that we did not choose *Output subroutines for part instances and feature pattern* as we did in Lesson 7 for outputting the G-code (see Figure 7.19 of Lesson 7). We expect that the G-code generated does not include subprogram calls. Therefore, the main program includes NC blocks that perform the nine machining operations four times for the respective four setups following the order shown in Figure 8.18.

To understand the G-code, we first locate the center points of a few selected holes. We use the *Measure* option (*Tools > Evaluate > Measure*) of SOLIDWORKS and choose *Center to Center* option; see Figure 8.22(a).

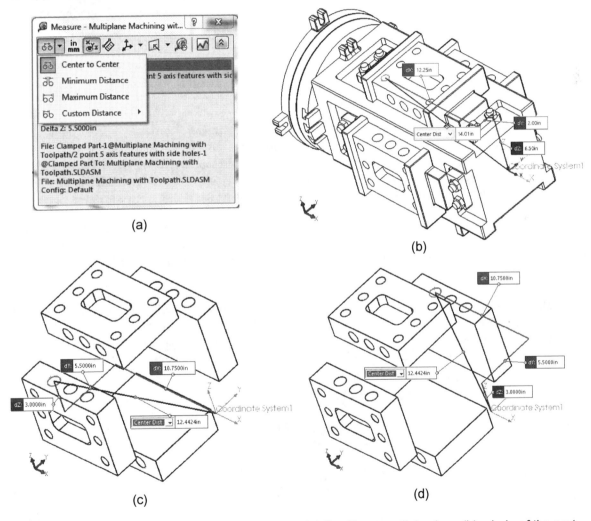

Figure 8.22 Coordinates of selected hole centers, (a) the *Measure* dialog box, (b) a hole of the part mounted on the top face of the tombstone selected, (c) a selected hole of the part mounted on the front side face of the tombstone (tombstone not shown), and (d) another hole of the part on the rear side face (tombstone not shown)

We first select the coordinate system (*Coordinate System1*) and pick the boundary edge of one of the holes of the part mounted on top; see Figure 8.22(b) for the hole selected. The coordinates of the center of the selected hole can be located at (–12.25, –2, 6.5), as shown in Figure 8.22(b). Since the distance between the neighboring holes is 2in., the coordinates of the center points of the remaining two holes close to the rear end of the top face are, respectively, (–12.25, 0, 6.5), and (–12.25, 2, 6.5).

Similarly, we select the coordinate system (*Coordinate System1*) and pick the boundary edge of one of the holes of the part mounted on the front side face; see Figure 8.22(c) for the hole picked. The coordinates of the hole center are (–10.75, –5.5, 3.0), as shown in Figure 8.22(c). Since the distance between the neighboring holes is 1.75in., the coordinates of the center points of the remaining two holes are, respectively, (–9.0, –5.5, 3.0), and (–7.25, –5.5, 3.0).

Also, we select the coordinate system (*Coordinate System1*) and pick the boundary edge of the hole of the part mounted on the rear side of the tombstone; see Figure 8.22(d). The coordinates of the hole center are (–10.75, 5.5, 3.0), as shown in Figure 8.22(d). Since the distance between the neighboring holes is 1.75in., the center points of the remaining holes are, respectively, (–9.00, 5.5, 3.0), and (–7.25, 5.5, 3.0).

You may click the *Post Process* button 🔲 above the graphics area, and follow the same steps learned in Lesson 4 to convert the toolpath into G-code. Figure 8.23 shows (partial) contents of the NC program (O0001).

As shown in Figure 8.23, the NC program is organized in four segments, led by identifier A defining the rotation angle of the 4th axis for the respective four setups. They are NC blocks *N7 A–90*, *N360 A–270*, *N713 A–180*, and *N1066 A0*, representing *Setup1* to *Setup4*, respectively. These rotation angles are consistent with those shown in Figure 8.18 except that a –90° angle is in place before *Setup1*. This is because the XZ plane was selected as zero degree position, as shown in Figure 8.8(a).

Each segment consists of nine operations. These operations, for example for *Setup1* (N7: A = –90), start at blocks *N6 (Rough Mill1)*, *N111 (Rough Mill2)*, *N230 (Contour Mill1)*, *N287 (Center Drill1)*, *N301 (Drill1)*, *N315 (Center Drill2)*, *N326 (Drill2)*, *N337 (Center Drill3)*, and *N348 (Drill3)*, for the respective operations.

We take a closer look at the *Center Drill1* operation. Blocks N288 and N289 move the center drill bit 7.5in. above the center point of the first hole (X = –12.25, Y = 2.0) on the top face of the part—the third hole to the right of the one selected in Figure 8.22(b). Block N290 makes a center drill to the hole, then to the next hole in the middle in N291 (X= –12.25, Y = 0), and then to the third hole in N292 (X = –12.25, Y = –2.0). This operation can also be seen by using the *Step Through Toolpath* option 🔲 , for example, blocks N287 to N300 for *Center Drill1*, as shown in Figure 8.24(a).

Also shown are blocks N315 to N325 for *Center Drill2* [Figure 8.24(b)], blocks N337 to N347 for *Center Drill3* [Figure 8.24(c)], and blocks N1043 to N1053 shown in Figure 8.24(d) for *Center Drill5* of *Setup3* with 4th axis: A–*180*.

The complete content of the G-code can be seen in the file: *Multiplane Machining no Subroutines.txt*.

We have now completed the lesson. You may save your model for future reference.

```
O0001
N1 G20
N2 (3/4 EM CRB 2FL 1-1/2 LOC)
N3 G91 G28 X0 Y0 Z0
N4 T04 M06
N5 S3677 M03

N6 ( Rough Mill1 )
N7 G90 G54 G00 X-10.265 Y.015 A-90. B0
N8 G43 Z6.6 H04 M08
N9 G01 Z6.125 F4.0448
N10 Y-.015 F16.1793
N11 X-8.235
N12 Y.015
N13 X-10.265
N14 Y.315
N15 X-10.565
N16 Y-.315
N17 X-7.935
N18 Y.315
N19 X-10.265
N20 Y.615
...

N111 ( Rough Mill2 )
N112 G90 G54 G00 X-10.75 Y.74
N113 G43 Z6.6 H03 M08
N114 G01 Z6.25 F6.8151
N115 G03 X-10.99 Y.5 I0 J-.24 F27.2606
N116 G01 Y.48
N117 G02 X-10.73 Y.74 I.26 J0
N118 G01 X-10.75
N119 G00 Z6.6
N120 Z6.35
N121 G01 Z6. F6.8151
...

N230 ( Contour Mill1 )
N231 G90 G54 G00 X-8.8837 Y.3273
N232 G43 Z6.6 H04 M08
N233 G01 Z6.125 F4.0448
N234 G41 D24 X-9.1595 Y.603 F12.1345
N235 G03 X-9.2125 Y.625 I-.053 J-.053
N236 G01 X-10.75 F16.1793
N237 G03 X-10.875 Y.5 I0 J-.125
N238 G01 Y-.5
...

N287 ( Center Drill1 )
N288 G90 G54 G00 X-12.25 Y2.
N289 G43 Z7.5 H17 M08
N290 G82 G98 R6.6 Z6.1443 P00 F26.4528
N291 Y0
N292 Y-2.
N293 X-6.25
N294 Y0
N295 Y2.
N296 G80 Z7.5 M09
N297 G91 G28 Z0
N298 (25/32 SCREW MACH DRILL)
N299 T18 M06
N300 S4967 M03

N301 ( Drill1 )
N302 G90 G54 G00 X-12.25 Y2.
N303 G43 Z7.5 H18 M08
N304 G83 G98 R6.6 Z5.5 Q.1 F29.8067
N305 Y0
...

N315 ( Center Drill2 )
N316 G90 G54 G00 X-10.75 Y-5.5
N317 G43 Z4. H19 M08
N318 G82 G98 R3.1 Z2.55 P00 F23.285
N319 X-9.
N320 X-7.25
N321 G80 Z4. M09
N322 G91 G28 Z0
N323 (1 SCREW MACH DRILL)
N324 T20 M06
N325 S3880 M03
```

```
N326 ( Drill2 )
N327 G90 G54 G00 X-10.75 Y-5.5
N328 G43 Z4. H20 M08
N329 G83 G98 R3.1 Z1.75 Q.1 F23.285
N330 X-9.
...

N337 ( Center Drill3 )
N338 G90 G54 G00 X-10.75 Y5.5
N339 G43 Z4. H19 M08
N340 G82 G98 R3.1 Z2.55 P00 F23.285
N341 X-9.
N342 X-7.25
N343 G80 Z4. M09
N344 G91 G28 Z0
N345 (1 SCREW MACH DRILL)
N346 T20 M06
N347 S3880 M03

N348 ( Drill3 )
N349 G90 G54 G00 X-10.75 Y5.5
N350 G43 Z4. H20 M08
N351 G83 G98 R3.1 Z1.75 Q.1 F23.285
N352 X-9.
...

N359 ( Rough Mill3 )
N360 G90 G54 G00 X-10.265 Y.015 A-270.
N361 G43 Z6.6 H04 M08
N362 G01 Z6.125 F4.0448
N363 Y-.015 F16.1793
N364 X-8.235
N365 Y.015
N366 X-10.265
N367 Y.315
N368 X-10.565
...

N464 ( Rough Mill4 )
N465 G90 G54 G00 X-10.75 Y.74
N466 G43 Z6.6 H03 M08
N467 G01 Z6.25 F6.8151
N468 G03 X-10.99 Y.5 I0 J-.24 F27.2606
N469 G01 Y.48
N470 G02 X-10.73 Y.74 I.26 J0
N471 G01 X-10.75
...

N583 ( Contour Mill2 )
N584 G90 G54 G00 X-8.8837 Y.3273
N585 G43 Z6.6 H04 M08
N586 G01 Z6.125 F4.0448
N587 G41 D24 X-9.1595 Y.603 F12.1345
N588 G03 X-9.2125 Y.625 I-.053 J-.053
N589 G01 X-10.75 F16.1793
...

N668 ( Center Drill5 )
N669 G90 G54 G00 X-10.75 Y-5.5
N670 G43 Z4. H20 M08
N671 G82 G98 R3.1 Z2.55 P00 F23.285
N672 X-9.
N673 X-7.25
N674 G80 Z4. M09
N675 G91 G28 Z0
N676 (1 SCREW MACH DRILL)
N677 T21 M06
N678 S3880 M03
...

N701 ( Drill6 )
N702 G90 G54 G00 X-10.75 Y5.5
N703 G43 Z4. H20 M08
N704 G83 G98 R3.1 Z1.75 Q.1 F23.285
N705 X-9.
N706 X-7.25
N707 G80 Z10. M09
N708 G91 G28 Z0
N709 (3/4 EM CRB 2FL 1-1/2 LOC)
N710 T04 M06
N711 S3677 M03
```

```
N712 ( Rough Mill5 )
N713 G90 G54 G00 X-10.265 Y.015 A-180.
N714 G43 Z6.6 H04 M08
N715 G01 Z6.125 F4.0448
N716 Y-.015 F16.1793
N717 X-8.235
N718 Y.015
N719 X-10.265
N720 Y.315
...

N1043 ( Center Drill9 )
N1044 G90 G54 G00 X-10.75 Y5.5
N1045 G43 Z4. H20 M08
N1046 G82 G98 R3.1 Z2.55 P00 F23.285
N1047 X-9.
N1048 X-7.25
N1049 G80 Z4. M09
N1050 G91 G28 Z0
N1051 (1 SCREW MACH DRILL)
N1052 T21 M06
N1053 S3880 M03 ...

N1054 ( Drill9 )
N1055 G90 G54 G00 X-10.75 Y5.5
N1056 G43 Z4. H20 M08
N1057 G83 G98 R3.1 Z1.75 Q.1 F23.285
N1058 X-9.
N1059 X-7.25
N1060 G80 Z10. M09
N1061 G91 G28 Z0
N1062 (3/4 EM CRB 2FL 1-1/2 LOC)
N1063 T04 M06
N1064 S3677 M03

N1065 ( Rough Mill7 )
N1066 G90 G54 G00 X-10.265 Y.015 A0
N1067 G43 Z6.6 H04 M08
N1068 G01 Z6.125 F4.0448
N1069 Y-.015 F16.1793
N1070 X-8.235
N1071 Y.015
N1072 X-10.265
N1073 Y.315
...

N1170 ( Rough Mill8 )
N1171 G90 G54 G00 X-10.75 Y.74
N1172 G43 Z6.6 H03 M08
N1173 G01 Z6.25 F6.8151
N1174 G03 X-10.99 Y.5 I0 J-.24 F27.2606
N1175 G01 Y.48
N1176 G02 X-10.73 Y.74 I.26 J0
N1177 G01 X-10.75
N1178 G00 Z6.6
N1179 Z6.35
...

N1396 ( Center Drill12 )
N1397 G90 G54 G00 X-10.75 Y5.5
N1398 G43 Z4. H19 M08
N1399 G82 G98 R3.1 Z2.55 P00 F23.285
N1400 X-9.
N1401 X-7.25
N1402 G80 Z4. M09
N1403 G91 G28 Z0
N1404 (1 SCREW MACH DRILL)
N1405 T20 M06
N1406 S3880 M03

N1407 ( Drill12 )
N1408 G90 G54 G00 X-10.75 Y5.5
N1409 G43 Z4. H20 M08
N1410 G83 G98 R3.1 Z1.75 Q.1 F23.285
N1411 X-9.
N1412 X-7.25
N1413 G80 Z4. M09
N1414 G91 G28 Z0
N1415 G28 X0 Y0
N1416 M30
```

Figure 8.23 The G-code generated by SOLIDWORKS CAM (partial contents)

8.6 Exercises

Problem 8.1. Create an assembly like that of Figure 8.25 using the parts (*Rotary Table*, *Problem 8.1 Part*, *Clamp 2*, and *Bolt half inch*) and subassembly (*Clamped Part*) in the Problem 8.1 folder downloaded from the publisher's website. Note that the freeform surface of *Problem 8.1 Part* is similar to that of Lesson 6. In addition to the freeform surface, there are holes at the top and side faces of the part. Create a total of four instances as a circular pattern feature.

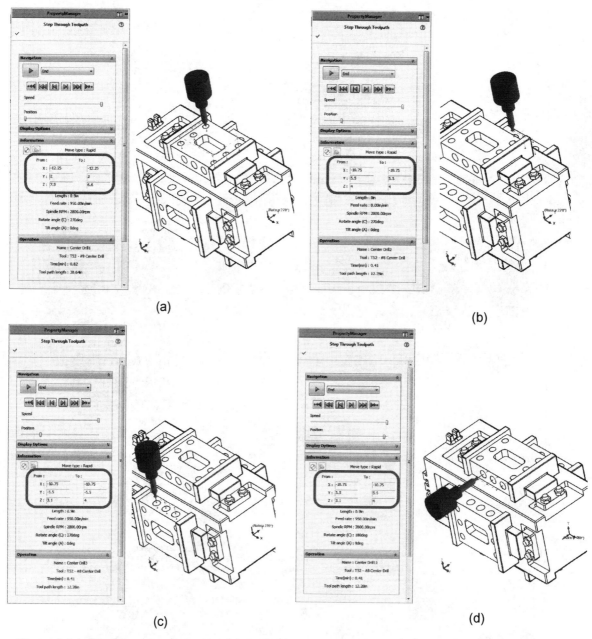

(a)

(b)

(c)

(d)

Figure 8.24 Step through the toolpath: (a) N287 to N300 (Center Drill1), (b) starting blocks N315 (Center Drill2), (c) starting blocks N337 (Center Drill3), and (d) starting blocks N1043 (Center Drill9, *Setup3*)

(a)   Generate machining operations for the four instances following steps discussed in this lesson. The operations must cut the freeform surface and holes on the top and side faces. Note that like Lesson 6, you will have to manually create a multi-surface machinable feature to machine the freeform surface.

(b)   Generate G-code and verify that the codes are generated correctly by reviewing the contents of the codes, similar to those of Section 8.5.

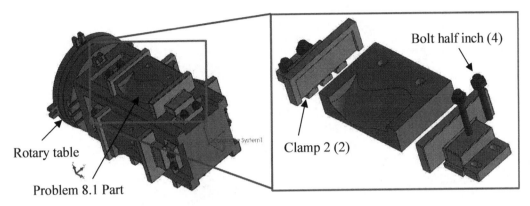

Figure 8.25 Assembly of Problem 8.1

Problem 8.2. Go back to the example in this lesson and select the *Output subroutines for part instances and feature patterns* option in the *Machine* dialog box (see Figure 8.26, and also see the discussion on Figure 7.19 of Lesson 7). Then, output G-code. Does the selection output G-code with subprogram calls in machining the instances? Verify if the G-code output from SOLIDWORKS CAM are correct.

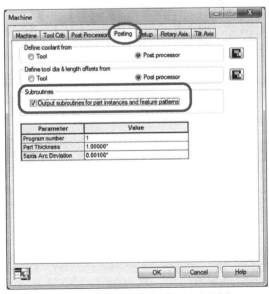

Figure 8.26 The *Posting* tab of the
Machine dialog box

[Notes]

Lesson 9: Tolerance-Based Machining

9.1   Overview of the Lesson

In this lesson, we learn tolerance-based machining (TBM). TBM is a plug-in for SOLIDWORKS CAM. This plug-in leverages SOLIDWORKS dimensions, tolerance ranges and surface finish annotations to select machining strategies for operations and machine parts to the mean of asymmetric tolerances. TBM automatically adjusts asymmetric tolerances to mean tolerances with which the machining strategies are created. This machining to mean capability eliminates long-standing issues surrounding differences between design practices required to tolerance parts based on fit, form and function versus manufacturing's need to machine geometry based on mean dimensions and tolerances. This capability reduces CNC programming time and enables you to efficiently carry out machining for parts meeting required tolerances. TBM supports 2.5 axis mill features as well as multi-surface mill features for the 2023 version of SOLIDWORKS CAM.

As we learned in previous lessons, most 2.5 axis features, such as pockets, holes, slots, and bosses, can be extracted as machinable features by using the automatic feature recognition (AFR) capability of SOLIDWORKS CAM. Also, machining strategies are automatically assigned to individual machinable features for machining operations as defined in the Technology Database. For example, the default machining strategy, drill, for hole drilling leads to two operations: center drill and drill.

Default machining strategy for individual machinable features can be reassigned to meet numerous tolerance requirements. For example, in addition to the center drill and drill operations for machining a hole, you may add machining operations by specifying machining strategy for a specific tolerance range, such as reaming and boring to meet prescribed tolerance requirements.

To enable TBM, the 3D part model designed using SOLIDWORKS must include product manufacturing information (PMI) data [1], which will have to be added to the solid model using *DimXpert* of SOLIDWORKS. The *DimXpert* tools within SOLIDWORKS are used to add details to the model for manufacturing by specifying tolerance features and associated 3D annotations. These 3D annotations (datums, dimensions, and geometric tolerances) are used to partially or fully document the part geometry.

The part solid model (filename: *block.SLDPRT*) of this lesson is similar to that of Lesson 5 (filename: *2 point 5 axis features.SLDPRT*), except that two additional pockets—Pockets 2 and 3, as shown in Figure 9.1(a)—are added. Overall, there are six holes and three pockets, all are 2.5 axis features. These features can be extracted as machinable features automatically. Also note that a 116° angle tip is added to the bottom face of the six blind holes. This revision is to alleviate a software issue in SOLIDWORKS CAM 2023 version, in which TMB-based machining strategy is not properly assigned to blind holes with flat bottom face.

---

1 Product and manufacturing information, also abbreviated PMI, conveys non-geometric attributes in 3D CAD for manufacturing product components and assemblies. PMI may include geometric dimensions and tolerances, 3D annotation and dimensions, surface finish, and material specifications. PMI is used in conjunction with the 3D model within model-based definition to allow for the elimination of 2D drawings for data set utilization.

Two solid models are provided for this lesson as a starting point. They are *block.SLDPRT*, which does not include any tolerance information; and *block for TBM.SLDPRT*, which has the tolerance information added by using *DimXpert* and is ready for you to start learning TBM directly. You may choose to learn the steps of using *DimXpert* to add tolerances to a solid model. In this case, you start with the block model without tolerance, i.e., *block.SLDPRT*. If you are already familiar with *DimXpert* or opt not to be sidetracked, you may skip Section 9.3 and directly open *block for TBM.SLDPRT* to go over this exercise.

In addition, two solid models with complete machining simulation are provided for your references. They are *Block with Toolpath.SLDPRT*, which includes toolpaths of a solid model without tolerances; and *Block for TBM with Toolpath.SLDPRT*, which has toolpaths created by using TBM. You may open these two models, review NC operations, and compare the toolpaths created for the same part with and without tolerances.

After completing this lesson, you should be able to add tolerances to a solid model using *DimXpert* (if you choose to go over Section 9.3), define tolerance ranges and respective machining strategies for TBM, and carry out machining operations for TBM.

9.2   The Machining Examples

As mentioned above, the block example includes three pockets and six holes. They are Pocket 1 (4.5in.×1.25in.×1.0in. with fillet radius 0.375in.), Pocket 2 (4in.×2in.×1.0in. with fillet radius 0.5in.), Pocket 3 (2.5in.×1.25in.×1.0in. with fillet radius 0.25in.), and holes of diameter 0.75in. and depth 1.0in. A stock of low carbon alloy steel 1005 with the size of the bounding box of the part is assumed; i.e., 8in.×6in.×2in. Note that unlike that of Lesson 5, no face milling is included for this lesson.

Block with Toolpath.SLDPRT

Machining operations for the block example of no tolerances (*Block.SLDPRT*) are generated by SOLIDWORKS CAM using the machining strategies defined in the TechDB™. The default strategy for machining a rectangular pocket is Rough-Rough(Rest)-Finish, leading to a rough mill, a rough rest mill, and a contour mill. For example, three operations generated for Pocket 2 (the center pocket) include Rough Mill3[T04 – 0.75 Flat End], Rough Mill4[T03 – 0.5 Flat End], and Contour Mill2[T04 – 0.75 Flat End], as shown in Figure 9.2.

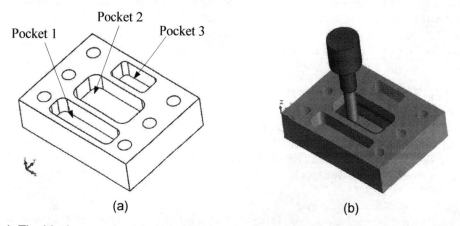

(a)   (b)

Figure 9.1 The block example, (a) part solid model with three rectangular pockets and six holes, and (b) material removal simulation

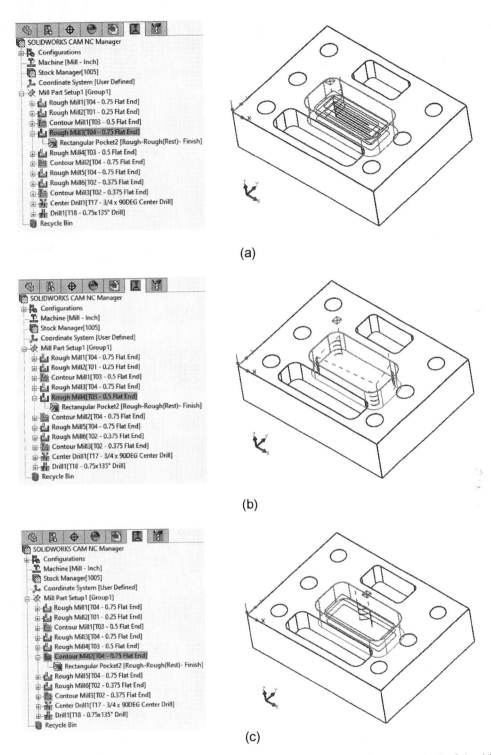

Figure 9.2 The machining operations generated for Pocket 2 (the center pocket) of the block example of no tolerance, (a) Rough Mill3[T04 – 0.75 Flat End], (b) Rough Mill4[T03 – 0.5 Flat End], and (c) Contour Mill2[T04 – 0.75 Flat End]

And the default strategy for drilling a hole leads to two operations: Center Drill1[T17 – 3/4×90DEG Center Drill], and Drill1[T18 – 0.75×135° Drill], as shown in Figure 9.3. Note that all NC operations and associated toolpaths were determined by SOLIDWORKS CAM.

Block for TBM with Toolpath.SLDPRT

Tolerances are defined for four dimensions of the solid part, filename: *Block for TBM.SLDPRT*, as shown in Figure 9.4. Three of them are for the length dimensions of the respective three pockets; i.e.,

Width_Pocket1: $4.500^{+.030}_{-.010}$ distance between the two opposite wall faces of Pocket 1.
Width_Pocket2: $4.00\pm.01$ distance between the two opposite wall faces of Pocket 2,
Width_Pocket3: $2.500^{+.000}_{-.001}$ distance between the two opposite wall faces of Pocket 3, and

Note that tolerance type of Width_Pocket2 (center pocket) is symmetric with feature dimension 4.00in. and tolerance limit ±0.01in.

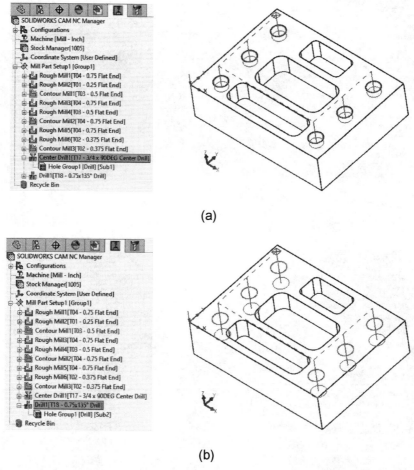

(a)

(b)

Figure 9.3 The machining operations generated for hole drilling of the block example of no tolerance, (a) Center Drill1[T17 – 3/4x90DEG Center Drill], and (b) Drill1[T18 – 0.75x135° Drill]

Width_Pocket1 and Width_Pocket3 are bilateral with different upper and lower tolerance limits. The feature dimension of Width_Pocket1 is 4.500in. with upper tolerance limit +0.030in. and lower tolerance limit −0.010in. Similarly, the feature dimension of Width_Pocket3 is 2.500in. with upper tolerance limit +0.000 and lower tolerance limit −0.001in.

The last tolerance is for the hole size, $\phi.750^{+.0050}_{-.0010} \triangledown 1.000$, in which the feature dimension of the diameter of the hole is 0.75in. with upper tolerance limit +0.005in. and lower tolerance limit −0.001in. No tolerance is defined for the depth of the hole.

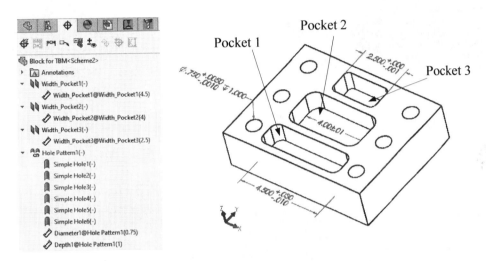

Figure 9.4 The tolerances defined for the four dimensions of the *block for TBM* example

Table 9.1 Machining strategies and operations generated for block with tolerance based on TBM and the same block of no TBM

Machinable Feature		Block for TBM with Toolpath (based on TBM)	Block for TBM with Toolpath (no TBM)
Pocket 1	Strategy	Rough-Finish-EdgeBreak	Rough-Rough(Rest)-Finish
	Machining Operations	Rough Mill1[T04 – 0.75 Flat End] Contour Mill1[T03 – 0.5 Flat End] Contour Mill2[T06 – 1/4×90 Countersink]	Rough Mill1[T04 – 0.75 Flat End] Rough Mill2[T01 – 0.25 Flat End] Contour Mill1[T03 – 0.5 Flat End]
Pocket 2	Strategy	Rough(VoluMill)-Rough(Rest)-Finish	Rough-Rough(Rest)-Finish
	Machining Operations	Rough Mill2[T04 – 0.75 Flat End] Rough Mill3[T03 – 0.5 Flat End] Contour Mill3[T04 – 0.75 Flat End]	Rough Mill3[T04 – 0.75 Flat End] Rough Mill4[T03 – 0.5 Flat End] Contour Mill2[T04 – 0.5 Flat End]
Pocket 3	Strategy	Rough	Rough-Rough(Rest)-Finish
	Machining Operations	Rough Mill4[T04 – 0.75 Flat End]	Rough Mill5[T04 – 0.75 Flat End] Rough Mill6[T02 – 0.375 Flat End] Contour Mill3[T02 – 0.375 Flat End]
Holes	Strategy	Ream	Drill
	Machining Operations	Center Drill1[T17–3/4×90DEG Center Drill] Drill1[T18–0.7344×135° Drill] Ream1[T19 – 0.75 Ream]	Center Drill1[T17–3/4×90DEG Center Drill] Drill1[T18–0.75×135° Drill]

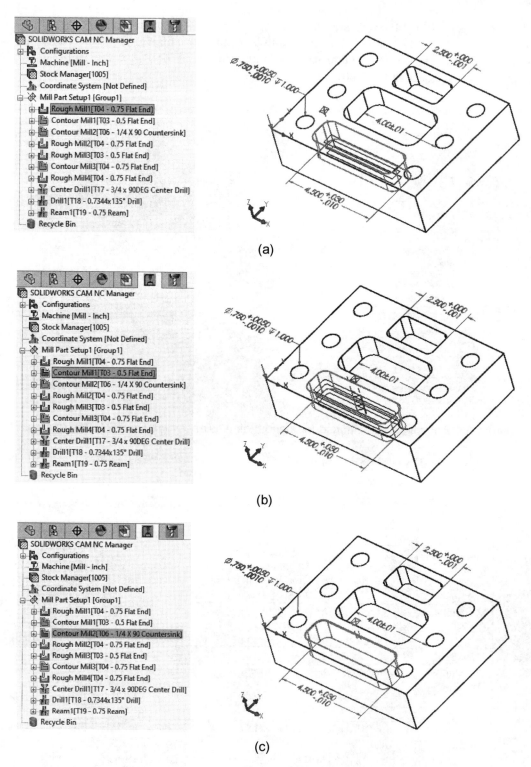

(a)

(b)

(c)

Figure 9.5 The machining operations generated for Pocket 1 of the block example, (a) Rough
Mill1[T04 – 0.75 Flat End], (b) Contour Mill1[T03 – 0.5 Flat End], and (c) Contour Mill2[T06 – 1/4×90
Countersink]

TBM settings, including the machining strategy assigned to the individual tolerance ranges of the respective machinable features (in this case, rectangular pockets and hole), are created for the block example with tolerances (filename: *Block for TBM with Toolpath.SLDPRT*). Machining operations are then generated by SOLIDWORKS CAM using the machining strategies defined for the respective tolerance ranges. A detailed discussion on the settings and tolerance ranges will be discussed in Section 9.4.

The strategies for machining the respective three pockets of the block with tolerances based on TBM are Rough-Finish-EdgeBreak, Rough(VoluMill)-Rough(Rest)-Finish, and Rough, as listed in Table 9.1 (also see Section 9.4 for details). The strategy for the hole drilling is Ream, as shown in Table 9.1. Also listed in Table 9.1 are the machining operations generated by running TBM based on the TBM settings.

For example, Rough-Finish-EdgeBreak is assigned to machining Pocket 1 in three operations, Rough Mill1[T04 – 0.75 Flat End], Contour Mill1[T03 – 0.5 Flat End], and Contour Mill2[T06 – 1/4×90 Countersink], as shown in Figure 9.5. Also, a Ream strategy is assigned to the six holes by running TBM. As a result, in addition to a center drill and a drill operation, a ream operation, Ream1[T19 – 0.75 Ream], are added to machine the holes. The toolpaths of the Ream1 operation are shown in Figure 9.6. More about running TBM based on the TBM settings will be discussed in Section 9.4.

In addition, the machining strategies and operations for machining the same block without tolerance information are listed in Table 9.1 (the right most column) for comparison.

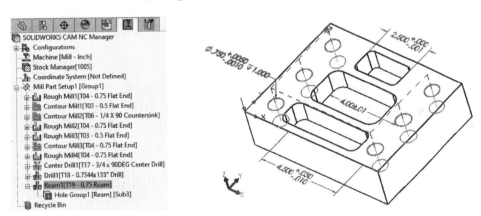

Figure 9.6 The last machining operation generated for hole drilling of the block example based on TBM, Ream1[T20 – 0.75 Ream]

9.3   Using *DimXpert for Parts* to Define Tolerances

DimXpert for parts is a set of tools you may use to apply dimensions and tolerances to parts according to the requirements of ASME Y14.41-2003[2] and ISO 16792:2006[3]. In this section, we illustrate the steps of

---

[2] ASME Y14.41 is a standard published by American Society of Mechanical Engineers (ASME) to establish requirements for model-based definition upon CAD software and those who use CAD software to create product definition within the 3D model. ASME issued the first version of this industrial standard on Aug 15, 2003, as ASME Y14.41-2003. It was immediately adopted by several industrial organizations, as well as the Department of Defense (DOD). ASME Y14.41 was revised and republished in May 2012 as ASME Y14.41-2012.

[3] ISO 16792:2006 specifies requirements for the preparation, revision and presentation of digital product definition data (data sets). It defines how to structure data and establishes digital product-definition data practices for both drawings and 3D models, either separately or in conjunction with one another. The overall goal of the standard is to foster consistent product-definition data practices and lay the groundwork for further technological developments. ASME Y14.41 served as the basis for the

using *DimXpert for parts* to create the four tolerance dimensions for the solid model *Block for TBM.SLDPRT*. If you are already familiar with *DimXpert*, you may skip this section and move directly to Section 9.4.

Open the part model *Block.SLDPRT*. Click the *MBD Dimensions* tab and choose the *Size Dimension* button , circled in Figure 9.7. Pick a wall face of Pocket 1, like that of Figure 9.8(a), and click the *Create Width Feature* button [circled in Figure 9.8(a)]. Rotate the view of the part like that of Figure 9.8(b) and pick the opposite side face of Pocket 1. Both faces are now selected and listed in the box. Click the checkmark ✓ to accept the selection, and place the dimension by clicking in the graphics window next to the pocket.

Figure 9.7 The *DimXpert* tab and the *Size Dimension* button

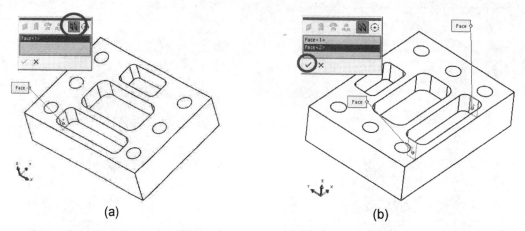

(a)   (b)

Figure 9.8 Picking side faces of Pocket 1 to define a size tolerance dimension, (a) picking the left side face and clicking the *Create Width Feature* button (circled), and (b) rotating the view and picking the opposite side face of the pocket

In the *DimXpert* dialog box—see Figure 9.9(a), choose *Bilateral* for *Tolerance/Precision* (since the upper and lower limits of the respective dimension tolerance are different.), enter *0.03in.* for the upper limit (+) and *–0.01in.* for the lower limit (–), as circled in Figure 9.9(a). You may adjust how the tolerance dimension is displayed in the graphics area. For example, you may deselect *Show parentheses* by clicking the checkbox in front of it—circled in Figure 9.9(b)—to show the tolerance value without parentheses. You may also adjust number of digits by clicking the drop-down button of the *Tolerance Precision* field and selecting, for example, *.123* for three digits; see Figure 9.9 (b). Click the checkmark ✓ to accept the tolerance dimension, a node *Width1* is listed in the *DimXpert* tree, as shown in Figure 9.10(a).

---

international standard *ISO 16792:2006 Technical product documentation -- Digital product definition data practices.* Both standards focus on the presentation of Geometric dimensioning and tolerancing (GD&T) together with the geometry of the product.

Click the node *Width1* and change its name to *Width_Pocket1*. Expand *Width_Pocket1* and change the name of the sub-node to *Width_Pocket1* as well, as shown in Figure 9.10(b). Repeat the same steps to create two width tolerance dimensions as follows:

Width_Pocket2: 4.00±.01 distance between the two opposite wall faces of Pocket 2, and

Width_Pocket3: 2.500$^{+.000}_{-.001}$ distance between the two opposite wall faces of Pocket 3.

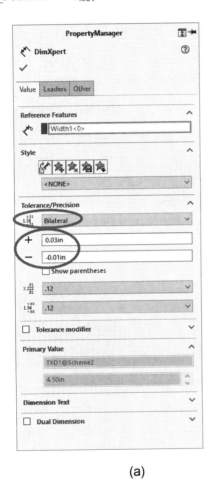
(a)

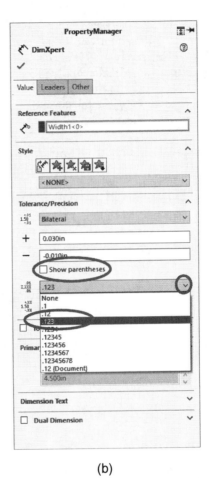
(b)

Figure 9.9 Defining the size tolerance dimension *Width1*, (a) choose *Bilateral* for *Tolerance/Precision*, enter *0.03in.* for the upper limit and *–0.01in.* for the lower limit, and (b) select number of digits for displaying the tolerance dimension

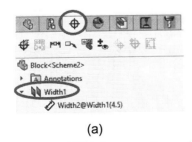
(a)

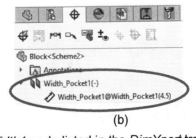
(b)

Figure 9.10 The width tolerance dimension, (a) *Width1* node listed in the *DimXpert* tree, and (b) tolerance dimension names changed to *Width_Pocket1*

Note that you will need to choose *Symmetry* for *Tolerance* type in the *DimXpert* dialog box for defining the Width_Pocket2 tolerance dimensions.

Next, we define hole tolerance dimension. Click the *DimXpert* tab and choose the *Size Dimension* button ⬛. Pick one of the six holes (the pattern feature button 🔳 will be selected automatically) and place the dimension by clicking in the graphics window close to the hole; see Figure 9.11(a).

In the *DimXpert* dialog box—see Figure 9.11(b), choose *Bilateral* for *Tolerance/Precision*, enter *0.005* and *–0.001* for the upper and lower limits (+ and –), respectively. Choose *HoleDepth* for *Callout value*, and choose *None* for *Tolerance/Precision* [see Figure 9.11(c)]. Click the checkmark ✓ to accept the tolerance dimension. A new node *Hole Pattern1* is now listed in the *DimXpert* tree.

We have completed defining the tolerance dimensions. You may save the part under a different filename before moving on to Section 9.4.

9.4   Tolerance-Based Machining Settings

The most important step in carrying out a successful TBM is defining the TBM settings. In this section, we discuss the TBM settings created for the two machinable features extracted: rectangular pocket and hole. These settings can be found in the part file: *Block for TBM with Toolpath.SLDPRT*, as shown in Figure 9.12(a) and Figure 9.12(b) for the rectangular pocket and hole, respectively. Detailed steps in creating the setups are provided in Section 9.5.

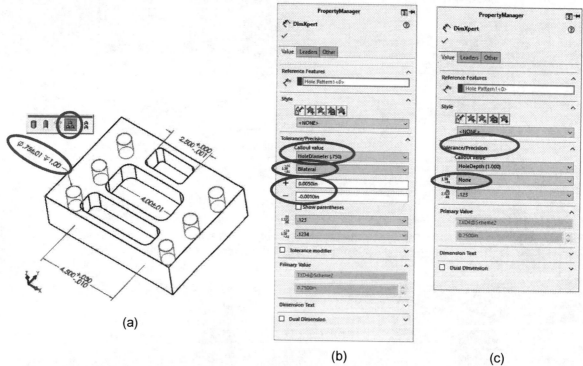

Figure 9.11 Defining the size tolerance dimension *Hole Pattern1*, (a) pick one of the six holes and choose the *pattern feature* button 🔳, (b) choose *Bilateral* for *Tolerance/Precision*, enter *0.005in.* for the upper limit and *–0.001in.* for the lower limit, and (c) select *None* for the depth dimension

You may open the model, *Block for TBM with Toolpath.SLDPRT*, choose *SOLIDWORKS CAM TBM* tab above the graphics area and click *Tolerance Based Machining (Mill) Settings* button to bring out the *Tolerance Based Machining (Mill) - Settings* dialog box.

In the *Tolerance Based Machining(Milling) – Settings* dialog box, choose the *Tolerance Range Mill (inch)* tab on top, and select, for example, *Rectangular Pocket* from the *Table of Features and Default Strategies* [the upper half of the dialog box shown in Figure 9.12(a)], and the tolerance ranges and respective machining strategies appear in the *Tolerance Based Condition grids* (lower half of the dialog box).

Table of Features and Default Strategies

There are three columns in the *Table of Features and Default Strategies* area (the upper half of the *Tolerance Based Machining(Milling) – Settings* dialog box). They are *Feature, Default Strategy*, and *Tol based conditions*. The *Feature* column (left most) lists 2.5-axis mill features for which tolerance based conditions have been defined. If the tolerance based conditions have not been defined for a particular feature type, then the font color for that feature type is shown in magenta color. If the tolerance based conditions have been completely defined for a particular feature type, then the font will be in black color.

For every feature type listed in the table of the *Tolerance Range Mill (mm)* and *Tolerance Range Mill (inch)* tabs, the corresponding entry in the *Default Strategy* field indicates the default machining strategy assigned to the feature determined by the TechDB™. For example, Rough-Rough(Rest)-Finish is listed as the default strategy for *Rectangular Pocket*.

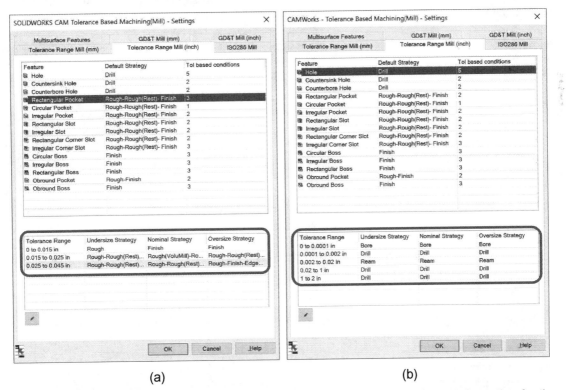

(a)   (b)

Figure 9.12 The *Tolerance Based Machining(Milling) – Settings* dialog box, (a) information for the *Rectangular Pocket* displayed in the *Tolerance Based Conditions grid* (lower half of the dialog box), and (b) information for the *Hole* displayed in the *Tolerance Based Conditions grid*

The corresponding entry in the *No. of Tol based conditions* column indicates the number of tolerance ranges defined for that particular feature type. For example, *3* is listed under the *No. of Tol based conditions* column for the *Rectangular Pocket*, indicating that three tolerance ranges and associated machining strategies have been defined for the feature.

If no tolerance based conditions have been defined for a particular feature type (i.e., the *No. of Tol based conditions* field indicates the value of '0'), then the font color for that particular feature type's listing within this tab will be magenta. If defined, then the font will be in black color.

When a particular feature type is selected, the tolerance based conditions (indicating the tolerance range and associated machining strategies) for this selected feature will be displayed in the *Tolerance based conditions grid* at the lower half of this window. For example, the three tolerance based conditions displayed in the grid while clicking *Rectangular Pocket* can be seen in Figure 9.12(a). Similarly, Figure 9.12(b) displays the five conditions for *Hole*.

Tolerance Based Conditions grid

As shown in Figure 9.12(a) and (b), the tolerance based conditions for the selected feature displayed in the *Tolerance based conditions grid* at the lower half of the dialog box consists of four columns: *Tolerance Range, Undersize Strategy, Nominal Strategy*, and *Oversize Strategy*.

The tolerance range column (left most) indicates the tolerance ranges defined for the feature type selected. For each range, the lower and upper limit of the tolerance is displayed along with the units (mm or inches). For example, three tolerance ranges are defined for the rectangular pocket, 0 to 0.015 in, 0.015 to 0.025 in, and 0.025 to 0.045 in, as shown in Figure 9.12(a).

And oversize strategy will be applied when the mean tolerance computed for the feature has a positive value; for example, Width_Pocket1, in which the mean tolerance value is $(0.03–0.01)/2 = 0.01$. Since the tolerance range of Width_Pocket3 is $0.03–(–0.01) = 0.04$, which is between 0.025 and 0.045 in.; an oversize strategy *Rough-Finish-EdgeBreak* is activated to machine Pocket 1.

The nominal strategy will be applied when the mean tolerance value computed for the feature is "0"; for example, Width_Pocket2, in which the mean tolerance value is $(0.01–0.01)/2 = 0$. Since the tolerance range of Width_Pocket2 is $0.01–(–0.01) = 0.02$, which is between 0.015 and 0.025 in., a nominal strategy *Rough(VoluMill)-Rough(Rest)-Finish* is chosen to machine Pocket 1.

The undersize strategy will be applied when the mean tolerance computed for the feature has a negative value. For example, the mean tolerance value of dimension Width_Pocket3 is $(0.000–0.001)/2 = –0.0005$ (undersize), and the tolerance range is $0.000–(–0.001) = 0.001$, which is between 0 and 0.015 in.; hence an undersize strategy *Rough* is chosen to machine Pocket 3.

When you define a new tolerance range for a selected feature in the *Tolerance Range* tab, the default strategy defined for the feature in the TechDB™ will be assigned for *Undersize, Nominal* and *Oversize* of the tolerance range. You can change/edit the *Undersize, Nominal* or *Oversize* strategy for a tolerance range by simply clicking the desired *Strategy* field.

Clicking the *Strategy* field displays a dropdown list of all machining strategies defined in the TechDB™ for that particular feature type, including user defined strategies. You may select the desired strategy from this dropdown list to reassign the strategy for a tolerance range. We will go over some of the steps in the next section.

9.5   Carrying Out Tolerance-Based Machining

In this section, we illustrate detailed steps in defining tolerance ranges and associated machining strategies for both the rectangular pocket and hole like those show in Figure 9.12(a) and Figure 9.12(b) using the *Tolerance Based Machining(Milling) – Settings* dialog box. Thereafter, we run a TBM and review machining operations generated.

Defining Tolerance Ranges for Machinable Features

We open the part file *Block for TBM.SLDPRT* or the part file you saved in Section 9.3. After opening the part, you should see tolerances defined for four dimensions, as shown in Figure 9.4.

We first click the *SOLIDWORKS CAM TBM* tab above the graphics area and click *Tolerance Based Machining(Milling) - Settings* button ⚙ to bring out the *Tolerance Based Machining(Milling) – Settings* dialog box.

Click the *Tolerance Range Mill (inch)* tab; you should see the default settings like those of Figure 9.13(a). Click *Rectangular Pocket*; the default tolerance settings appear in the *Tolerance based conditions grid* at the lower half of this tab. Initially only two tolerance ranges are defined for rectangular pocket; i.e., *0 to 0.002in.*, and *0.002 to 1in*. Note that you may see different ranges as those of Figure 9.13(a) depending on the current settings of the software.

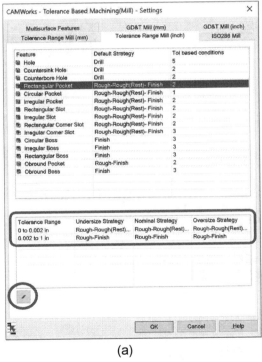

(a)   (b)

Figure 9.13 The *Tolerance Based Machining (Milling) – Settings* dialog box, (a) two tolerance ranges of the *Rectangular Pocket* displayed in the *Tolerance Based Conditions grid (lower half of the dialog box)*, (b) entering four numbers for the three tolerance ranges, and (c) three tolerance ranges listed in the *Tolerance Based Conditions grid*

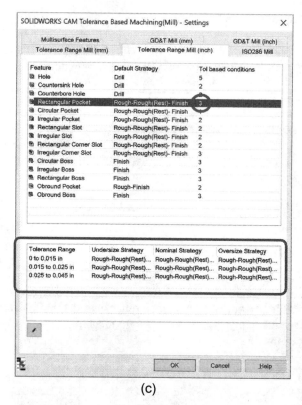

(c)

Figure 9.13 The *Tolerance Based Machining (Milling) – Settings* dialog box (cont'd)

In this lesson, the existing tolerance ranges will be replaced with the following three tolerance ranges: 0 to 0.015in., 0.015 to 0.025 in., and 0.025 to 0.045in. like those shown in Figure 9.12(a).

To edit the tolerance ranges, click the *Edit Tolerance Range* button ![pencil] at the bottom left corner of the *Tolerance Range* tab—circled in Figure 9.13(a). The *Range* dialog box appears; see Figure 9.13(b). Delete the range data by clicking a cell and pressing the *Delete* key one at a time.

In the text field to the left of the *Add* button ![plus], enter *0* and press the *Add* button; the number entered will appear in the first cell. Follow the same steps to enter *0.015*, *0.025*, and *0.045*. Click *OK* to accept the numbers entered.

Observe the table of *Features and Default Strategies* [that is, the upper half of Figure 9.13(c)]. In the *No. of Tol based Conditions* field for the rectangular pocket feature, the value has changed from *2* to *3*, circled in Figure 9.13(c), indicating that three tolerance ranges are now defined for the rectangular pocket feature.

For all these new tolerance ranges that have been defined, the machining strategy assigned to *Undersize Strategy*, *Nominal Strategy* and *Oversize Strategy* is the same as the default strategy for the rectangular pocket feature; i.e., *Rough-Rough(Rest)-Finish*—circled in Figure 9.13(c).

Reassigning Strategies for Tolerance Ranges

In this step, we assign desired machining strategies to the respective tolerance ranges for the rectangular pocket feature.

In the *Tolerance Based Conditions grid* [that is, the lower half of Figure 9.13(c)], click to highlight the first range: *0.0 to 0.015 in.* For this selected entry, click the *Undersize Strategy* field and select *Rough* from the dropdown list (see Figure 9.14). Similarly, click the *Nominal Strategy* field and select *Finish* from the dropdown list, and click the *Oversize Strategy* field and select *Finish* from the dropdown list.

In the *Tolerance Based Conditions grid*, click to highlight the second range: *0.015 to 0.025in.* and assign *Rough(VoluMill)-Rough(Rest)-Finish* to *Nominal Strategy* only.

Similarly, we choose the third tolerance range: *0.025 to 0.045in.* and assign *Rough-Finish-EdgeBreak* to *Oversize Strategy* only.

The machining strategies for the three tolerance ranges of the *Rectangular Pocket* are now identical to those shown in Figure 9.12(a).

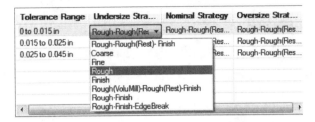

Figure 9.14 Assigning machining strategies

We follow the same steps to assign machining strategies for the five tolerance ranges of the hole machinable feature, as shown in Figure 9.12(b).

Click the *OK* button to accept the changes.

Carry Out Tolerance Based Machining

On the *SOLIDWORKS CAM TBM Command Manager* above the graphics area, click the *Tolerance Based Machining - Run* button ![icon] . The *Tolerance Based Machining (Mill)* – *Run* dialog box appears; see
Figure 9.15(a). The *Run* tab provides checkbox options to select the processes to be executed.

In this lesson, machining strategies have been assigned to rectangular pocket and hole; both are 2.5 axis features, based on tolerance ranges we have defined, as shown in Figure 9.12(a) and (b). Hence, we place a check in the checkbox labeled *Recognize tolerance range*.

Choose the *Tolerance Range Mill (inch)* tab. Currently, no condition has been identified for any of the machinable features before running the analysis; "0 of 0" is shown under the *Identified features* column and all entities are in magenta color, as shown in
Figure 9.15(b).

Under the *Tolerance Range Mill (inch)* tab of the *Tolerance Based Machining (Mill)* – *Run* dialog box, click the *Extract Machinable Features* button ![icon] —at lower left corner of the dialog box, as circled in
Figure 9.15(b)—to recognize machinable features.

Four features are recognized and are listed under *Mill Part Setup1* of the SOLIDWORKS CAM feature tree tab ![icon], as shown in Figure 9.16(a), with desired machining strategies assigned; for example, *Rough* is assigned to Pocket 3 (instead of the default *Rough-Rough(Rest)-Finish*). Also, *Hole* and *Rectangular Pocket* are turned into black color with, respectively, *1 of 1* and *3 of 3* conditions identified under the *Identified features* column, as shown in Figure 9.16(b). Click, for example, the *Rectangular Pocket* to show the tolerance ranges defined and machining strategies assigned in the lower half of the dialog box [see Figure 9.16(b)].

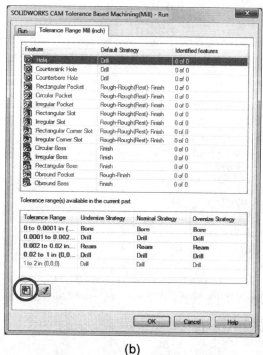

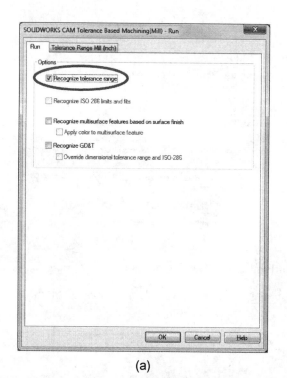

(a)   (b)

Figure 9.15 The *Tolerance Based Machining (Mill)* – *Run* dialog box, (a) selections under the *Run* tab, and (b) "0 of 0" listed under *Identified features* column, and all in magenta color under the *Tolerance Range (inch)* tab

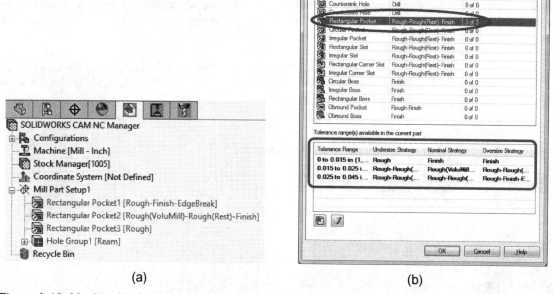

(a)   (b)

Figure 9.16 Machinable features recognized, (a) four features listed under *Mill Part Setup1*, and (b) "1 of 1" for *Hole* and "3 of 3" for *Rectangular Pocket* listed under *Identified features* column

Click the *OK* button to close this dialog box.

Right click *Mill Part Setup1* under the SOLIDWORKS CAM feature tree tab ![icon] and choose *Generate Operation Plan*. Eleven operations are generated, as listed under *Mill Part Setup1* of the SOLIDWORKS CAM operation tree tab ![icon]. They are in magenta color as shown in Figure 9.17(a).

Right click *Mill Part Setup1* under the SOLIDWORKS CAM operation tree tab ![icon] and choose *Generate Toolpath*. All toolpaths are generated.

We take a closer look at the results next.

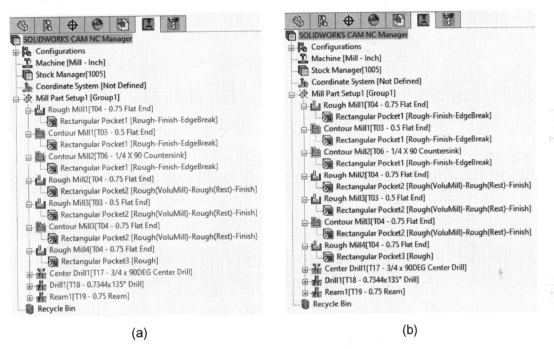

(a) (b)

Figure 9.17 Machining operation plans and toolpath generated, (a) ten operation plans generated for the respective four machinable features, and (b) machining toolpaths generated

9.6   Interpreting Information in the Tolerance Range Tab

On the *SOLIDWORKS CAM TBM Command Manager*, once again click the *Run Tolerance Based Machining* button to bring up the *Tolerance Based Machining (Mill) – Run* dialog box. Click the *Tolerance Range (inch)* tab in this dialog box.

Observe the table of listed features and machining strategies. Again, the last column in the table, *Identified features*, indicates whether matching tolerance ranges were found or not for the corresponding feature type on the part model.

If one or more matching tolerance based conditions have been identified for a 2.5 axis feature recognized on the solid part, then the corresponding entry for that feature in this table will be in black color font. The *Identified features* field will indicate the number of matching tolerance based conditions.

For this lesson, the entry for the rectangular pocket and hole features appears in black font color, and *3 of 3* and *1 of 1* are listed under the *Identified features* field for the rectangular pocket and hole, respectively; see Figure 9.18(a) and (b).

Click a feature type in black color to display tolerance based conditions listed in *Tolerance Based Conditions grid* (lower half of the dialog box). Adjacent to the tolerance range value are three integer numbers separated by commas and contained within parentheses. For the selected feature type in the *Tolerance Range* tab, the first, second, and third numbers indicate the number of feature instances present in the feature tree which fulfills the *Tolerance Based Condition* such that, respectively, the *Undersize Strategy*, *Nominal Strategy*, and *Oversize Strategy* for machining will be applied.

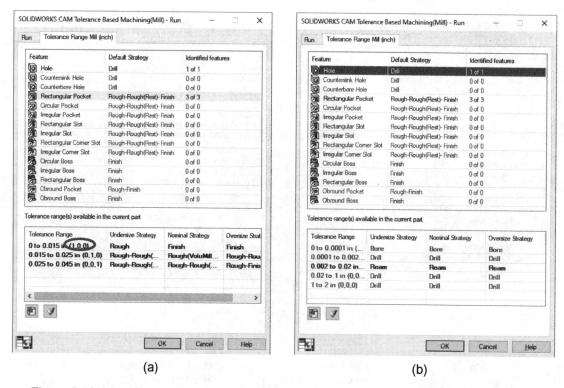

(a)   (b)

Figure 9.18 Identified conditions for the machinable features listed in the *Tolerance Based Machining (Mill) – Run* dialog box, (a) rectangular pocket, and (b) hole

For example, click the *Rectangular Pocket* row; the tolerance ranges appear in the *Tolerance Based Conditions grid*—see Figure 9.18(a). The integers in the parentheses under the *Tolerance Range* column are (1, 0, 0) for range 0 to 0.015in. [circled in Figure 9.18(a)], (0, 1, 0) for range 0.015 to 0.025in., and (0, 0, 1) for range 0.025 to 0.045in., respectively.

This is because that *Width_Pocket1* is oversize with tolerance range 0.04 in. (see Table 9.2), hence (0, 0, 1) and *Rough-Finish-EdgeBreak* is assigned as *Oversize Strategy* to machine Pocket 1. As a result, three operations, *Rough Mill1*, *Contour Mill1*, and *Contour Mill2*, are generated for machining Pocket 1.

Similarly, *Width_Pocket2* is nominal with tolerance range 0.02 in. (see Table 9.2), hence (0, 1, 0) and *Rough(Volu)-Rest-Finish* is assigned as *Nominal Strategy* to machine Pocket 2. As a result, three operations, *Rough Mill2*, *Rough Mill3*, and *Contour Mill3*, are generated for machining Pocket 2.

Also, *Width_Pocket3* is undersize with range 0.001 in. (see Table 9.2), hence (1, 0, 0) and *Rough* is assigned as *Undersize Strategy* to machine Pocket 3.

Click the *Hole* row; the tolerance ranges appear in the *Tolerance Based Conditions grid*—see Figure 9.18(b), in which only the third row of range 0.002 to 0.02in. is in black font. For the tolerance range, 0.002 to 0.02in., of the hole feature, the numbers in the parentheses are (0, 0, 1) implying that *Ream* is applied to the *Oversize Strategy* of the hole drilling. This is because, as shown in Table 9.2, the hole diameter has a mean tolerance of 0.002in., which is oversize, and tolerance range 0.006in., which falls into the 0.002 to 0.02in. range.

We have now completed the lesson. You may save your model for future reference.

Table 9.2 Comparison of machining strategies and operations between block of no tolerance and with tolerances

Tolerance Dimension	Mean Tolerance	Tolerance Range	Strategy
Width_Pocket1: $4.500^{+.030}_{-.010}$	0.01 (oversize)	0.04in.	Rough-Finish-EdgeBreak
Width_Pocket2: $4.00\pm.01$	0 (nominal)	0.02in.	Rough(VoluMill)-Rough(Rest)-Finish
Width_Pocket3: $2.500^{+.000}_{-.001}$	−0.0005 (undersize)	0.001in.	Rough
Hole Pattern1: $\varnothing.750^{+.0050}_{-.0010} \, \nabla 1.000$	0.002 (oversize)	0.006in.	Ream

9.7   Exercises

Problem 9.1. Generate machining operations to cut the part shown in Figure 9.19 from a stock of 4in.×4.25in.×1.5in. Use *DimXpert* to assign tolerances to the width dimension of the pocket. The feature dimension of the width is 2.00in. with upper tolerance limit +0.000 and lower tolerance limit −0.010in. Use the same tolerance settings as shown in Figure 9.18(a) to carry out a TBM simulation.

Submit a table similar to that of Table 9.1 that compares the machining operations for the same part with and without tolerance information. Also, submit screen captures as needed to demonstrate your work.

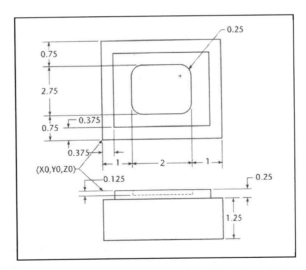

Figure 9.19 The design model of Problem 9.1

[Notes]

Lesson 10: Turning a Stepped Bar

10.1 Overview of the Lesson

We discuss turning operations in Lessons 10 and 11. In Lesson 10, we use a simple stepped bar example to learn basic capabilities in creating turning operations and understanding G-code post-processed by SOLIDWORKS CAM. In Lesson 11, we machine a similar bar with more turning features to gain a broader understanding of the turning capabilities offered by SOLIDWORKS CAM. Note that SOLIDWORKS CAM Standard version does not support turning operations. You will have to use the Professional version to go over this lesson.

This current lesson should provide you with a quick run-through in creating turning operations using SOLIDWORKS CAM. You will learn a complete process in using SOLIDWORKS CAM to create turning operations from the beginning all the way to the post process that generates G-code. We use a lathe of single turret to machine the simple stepped bar shown in Figure 10.1(a) from a round stock clamped into a three-jar chuck, as shown Figure 10.1(b). We create an outer profile turning operation (called outer diameter or OD turning in SOLIDWORKS CAM) and a cut off operation to remove the part from the stock. The machined part at the end of the material removal simulation is shown in Figure 10.1(c). This lesson is intentionally made simple. We stay with default options and machining parameter values for most of the lesson.

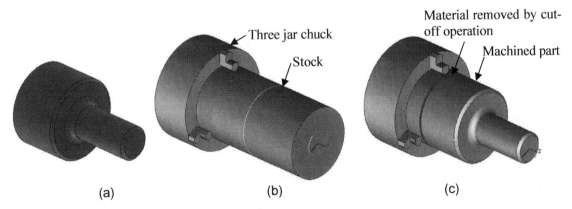

Figure 10.1 The stepped bar example, (a) the design model, (b) bar stock clamped to a chuck, and (c) the material removal simulation

Similar to the milling lessons, we follow the general steps shown in Figure 1.1 of Lesson 1 to turn the stepped bar. We will start by (1) opening the stepped bar design model; (2) defining machine setup, in which we choose a single turret lathe, select a tool crib, pick a post processor, and choose a fixture coordinate system; (3) creating a stock, in this case, a cylindrical bar enclosing the design model; (4) defining machinable features manually and later using automatic feature recognition (AFR) in Lesson 11; (5) generating operation plans and toolpaths; and (6) checking results and reviewing material removal simulation. In addition, we will go over a post process to create G-code of selected operations, including turn finish and cut off. We will take a closer look at the G-code generated to gain a better understanding of the turning operations.

After completing this lesson, you should be able to carry out machining simulation for similar turning operations following the same procedures and be ready to move onto Lesson 11.

10.2 The Stepped Bar Example

The stepped bar shown in Figure 10.2(a) has a bounding cylinder of size φ4.25in.×6.5in. The unit system chosen is IPS. There is one revolve feature created by revolving the sketch shown in Figure 10.2(b). In addition, there is a chamfer of 0.15in. defined at two boundary edges shown in Figure 10.2(a), and a fillet (0.375in. in radius) feature, plus a point and a coordinate system. These features are listed in the feature tree shown in Figure 10.3. The coordinate system (*Coordinate System1*) and the point (*Point1*) are chosen as the fixture coordinate system and the part setup origin, respectively, for turning operations. When you open the solid model *Stepped bar.SLDPRT*, you should see the solid and the reference features listed in the feature tree like those shown in Figure 10.3.

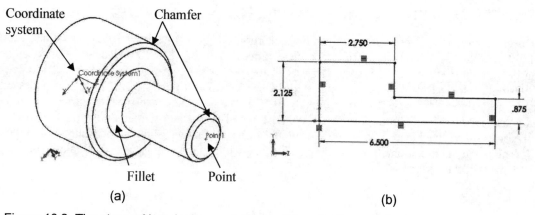

(a)   (b)

Figure 10.2 The stepped bar design model, (a) solid and reference features, and (b) sketch of the revolved solid feature

A bar stock of size φ4.25in.×7.75in., made of low carbon alloy steel (1005), as shown in Figure 10.4(a), is chosen for the turning operations. The front end of the stock coincides with the front face of the part. The rear end of the stock is clamped in the three-jar chuck with an extra length of 1.25in. to be cut off at the end of the turning operation. Since the front end face of the part coincides with that of the stock, there is no need to define a face turning operation.

Note that a part setup origin is defined at the center of the front end face of the stock (coinciding with *Point1* defined in part), which defines the G-code program zero location.

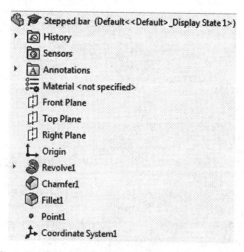

Figure 10.3 Entities listed in the feature tree

We will create two machinable features — an OD (outer diameter) feature that defines the outer shape of the part from the front end face to the cut off face, and a cut off feature that cuts off the part from the stock. In general, an OD machinable feature excludes the shape of any groove features (which are not present in this simple example).

We create these machinable features manually. In the next lesson, we will use the automatic feature recognition (AFR) capability to recognize machinable features from the part solid model. In this lesson, we follow the recommendations of the technology database (TechDB™) for choosing machining options, cutters, and cutting parameters.

The toolpaths of the turning operations for machining the OD feature, consisting of *Turn Rough* and *Turn Finish*, are shown in Figure 10.5(a) and Figure 10.5(b), respectively. The turn rough toolpath moves the cutter along the outer profile of the part in numerous passes. This is because the cut amount is chosen as 0.1in. and the overall depth to cut is 1.25in. The turn finish toolpath moves the cutter along the part outer boundary in one pass, similar to the contour mill operation in milling. The cut off toolpath simply moves the cutter along the negative X direction to separate the part from the stock, as shown in Figure 10.5(c).

You may open the example file with toolpath created (filename: *Stepped bar with toolpath.SLDPRT*) to preview its toolpath and machining simulation.

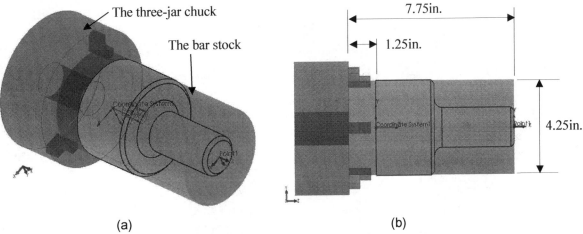

Figure 10.4 A bar stock of ⌀4.25in.×7.75in., (a) stock clamped into a chuck, and (b) stock size

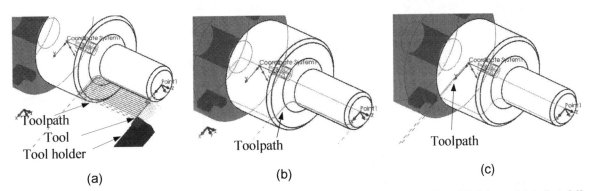

Figure 10.5 The toolpaths of the turning operations, (a) *Turn Rough*, (b) *Turn Finish*, and (c) *Cut Off*

10.3 Using SOLIDWORKS CAM

Open SOLIDWORKS Part

Open the stepped bar design model (filename: *Stepped bar.SLDPRT*) downloaded from the publisher's website. This solid model, as shown in Figure 10.3, consists of three solid features (revolve, chamfer, and fillet), a point, and a coordinate system. As soon as you open the model, you may want to check that the IPS unit system is chosen and increase the decimals from the default 2 to 4 digits.

Select NC Machine

Click the SOLIDWORKS CAM feature tree tab 🛠. Right click *Machine* and select *Edit Definition*. In the *Machine* dialog box (Figure 10.6), *Mill-inch* is selected under *Machine* tab. We choose *Turn Single Turret - Inch* from the list of *Applicable machines* box, and click *Select*.

Choose *Tool Crib* tab and select *Tool Crib 2 Rear* under *Available tool cribs* (Figure 10.7), and then click *Select*. Choose the *Post Processor* tab; a post processor called *T2AXIS-TUTORIAL* is selected by default (Figure 10.8). This is a generic post processor of 2-axis lathe that comes with SOLIDWORKS CAM. There are more post processors that come with SOLIDWORKS CAM; they are located in *C:\ProgramData\SOLIDWORKS\SOLIDWORKS CAM 2023\Posts* folder in your computer.

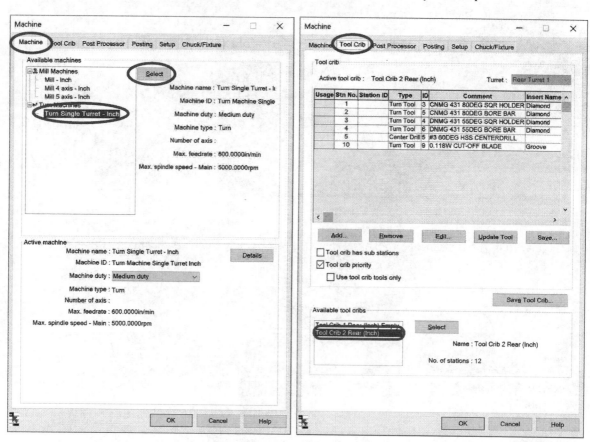

Figure 10.6 The *Machine* tab of the *Machine* dialog box

Figure 10.7 The *Tool Crib* tab of the *Machine* dialog box

Note that in practice you will need to identify a suitable post processor that produces G-code compatible with the CNC lathes on the shop floor. We will not make any change in the post processor selection for this lesson. Then, click the *Setup* tab, and click *User Defined* to choose *Coordinate System1* for *Main spindle coordinate system* (see Figure 10.9). Click *OK* to accept the selections and close the dialog box.

Create Stock

From SOLIDWORKS CAM feature tree 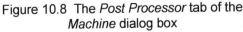, right click *Stock Manager* and choose *Edit Definition*. The *Stock Manager* dialog box appears (Figure 10.10), in which a default stock size appears under *Bar stock parameters*, including outer diameter ▦ (4.25in.), inner diameter ▦ (0in.), overall length ▦ (6.5in.), and back of stock absolute ▦ (0in.) that defines length of the stock outside the part.

We will increase the overall length of the stock to 7.75in. and enter the length of the stock outside the part to be −1.25in., as shown in Figure 10.10. We choose *Low Carbon Alloy Steel 1005* for stock material. Accept the revised stock by clicking the checkmark ✓ at the top left corner. The bar stock should appear in the graphics area similar to that of Figure 10.4.

Turn Setup and Machinable Feature

We create two machinable features manually, OD and cut off. We first create a turn setup and then insert new turn features under the SOLIDWORKS CAM feature tree.

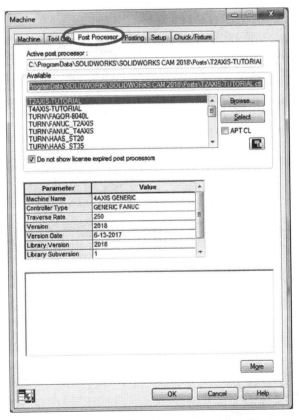

Figure 10.8 The *Post Processor* tab of the *Machine* dialog box

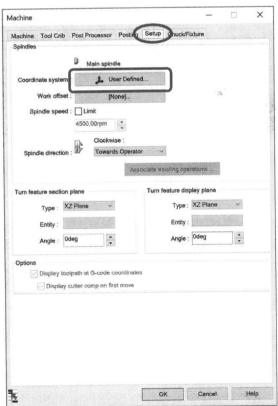

Figure 10.9 The *Setup* tab of the *Machine* dialog box

Under the SOLIDWORKS CAM feature tree tab 📐, right click *Stock Manager* and choose *Turn Setup*. The *Turn Setup* dialog box appears (Figure 10.11). In the graphics area, a coordinate system of the part setup appears at the rear end face of the part (circled in Figure 10.12), coinciding with the fixture coordinate system, *Coordinate System1*. This coordinate system indicates that the Z-axis aligns with the spindle direction, which is adequate.

We will relocate the part setup origin to the front end face of the stock later. For the time being, click the checkmark ✓ in the *Turn Setup* dialog box to accept the definition. A *Turn Setup1* is now listed in the SOLIDWORKS CAM feature tree 📐.

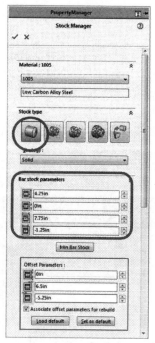

Figure 10.10 The *Stock Manager* dialog box

Figure 10.11 The *Turn Setup* dialog box

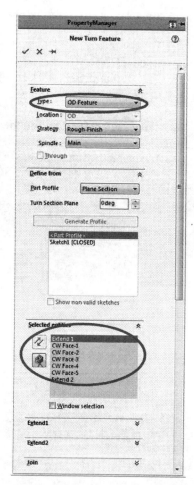

Figure 10.13 The *New Turn Feature* dialog box

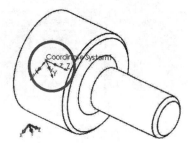

Figure 10.12 The turning coordinate system

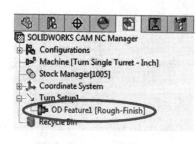

Figure 10.15

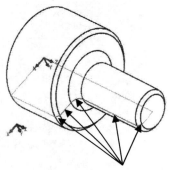

Pick these five edges

Figure 10.14 Picking these five edges to define an OD feature

Now we define machinable features. Under SOLIDWORKS CAM feature tree tab ![icon], right click *Turn Setup1* and choose *Turn Feature*. In the *New Turn Feature* dialog box (Figure 10.13), choose *OD Feature* for *Type*, and pick the five edges that define the boundary profile of the OD feature in the graphics area, as shown in Figure 10.14. Click the checkmark ✔ in the *New Turn Feature* dialog box to accept the definition.

An *OD Feature1* node is now listed in the SOLIDWORKS CAM feature tree ![icon] in magenta color (see Figure 10.15).

We follow the same steps to create a cut off machinable feature; i.e., right click *Turn Setup1* and choose *Turn Feature*. In the *New Turn Feature* dialog box, choose *CutOff Feature* for *Type*, and pick the edge at the rear end face of the part in the graphics area, as shown in Figure 10.16. Click the checkmark ✔ in the *New Turn Feature* dialog box to accept the definition.

A *CutOff Feature1* node is now listed in the SOLIDWORKS CAM feature tree ![icon] in magenta color.

Generate Operation Plan and Toolpath

Right click *Turn Setup1* and choose *Generate Operation Plan* (or click the *Generate Operation Plan* button ![icon] above the graphics area).

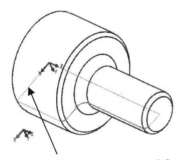

Pick this edge at the rear end face

Figure 10.16 Picking this edge to define a cutoff feature

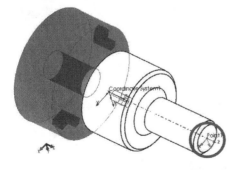

Figure 10.18 The part setup origin relocated to the center of front end face

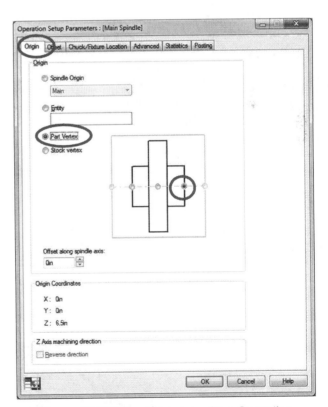

Figure 10.17 The *Origin* tab of the *Operation Setup Parameters* dialog box

Three operations, *Turn Rough1*, *Turn Finish1*, and *Cut Off1*, are listed in SOLIDWORKS CAM operation tree [⬛] in magenta color. Right click *Turn Setup1* and choose *Generate Toolpath* (or click the *Generate Toolpath* button [Generate Toolpath] above the graphics area). Turning toolpaths will be generated like those shown in Figure 10.5.

Relocate Part Setup Origin

A part setup origin has been chosen by SOLIDWORKS CAM to be coincident with the fixture coordinate system, *Coordinate System1* (see Figure 10.16). We will move the part setup origin to *Point1* (located at the center of the front end face of the part), which is more practical since this point is easier to access in a stock clamped into a chuck on a lathe.

Under the SOLIDWORKS CAM operation tree tab [⬛], right click *Turn Setup1* and choose *Edit Definition*.

In the *Operation Setup Parameters* dialog box, choose *Origin* tab, and select *Part Vertex*. The vertex at the right end of the sample part in the sketch (circled in Figure 10.17) is selected, which matches our intent. The part setup origin moved to the center of the front end face (see Figure 10.18) as desired. Click *OK* in the *Operation Setup Parameters* dialog box to accept the change. Click *Yes* to the question in the warning box: *The origin or chuck/fixture location/avoidance parameters has changed, toolpaths need to be recalculated. Regenerate toolpaths now?*

Next we review the tool and key machining parameters of the first turning operation, *Turn Rough1*.

Review the Operations

Under the SOLIDWORKS CAM operation tree tab [⬛], right click *Turn Rough1* and choose *Edit Definition*. In the *Operation Parameters* dialog box, choose *Diamond Insert* tab; a diamond insert of $0.0157 \times 80°$ (radius 0.0157in., angle: 80°, and inscribed circle 0.5in., ID:1, CNMG 431) appears.

Change the insert radius to 0.02 (circled in Figure 10.19). Click the *Holder* tab (see Figure 10.20). A standard holder with a shank width 0.75in. and length 4in. (Holder ID:1, RH 80DEG SQR HOLDER) has been chosen. Also, the *Down left* is chosen for *Orientation*, which defines the orientation of the cutter suitable to turn the OD feature.

Choose the *Turn Rough* tab of the *Operation Parameters* dialog box. In the *Profile parameters* area, the *First cut amt.* and *Max cut amt.* are set to *0.1in*, and *Final cut amt.* is *0.025in.*, as shown in Figure 10.22. These parameters define the distance of the tool movement along the negative X-direction, similar to the depth of cut in milling. Click *OK* to accept the selections and close the dialog box. You may review other operations following the same steps.

Note that we also change the insert radius to 0.02 for the *Turn Finish1* operation as well. This is to simplify a bit for reviewing the toolpath next.

Simulate Toolpath

Right click *Turn Setup1* and choose *Simulate Toolpath* (or click the *Simulate Toolpath* button [Simulate Toolpath] above the graphics area) to the simulate toolpath. The material removal simulation appears similar to that of Figure 10.1(c).

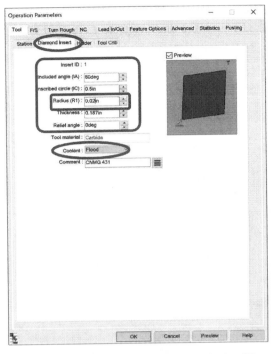

Figure 10.19 The *Diamond Insert* tab of the *Tool* tab in the *Operation Parameters* dialog box

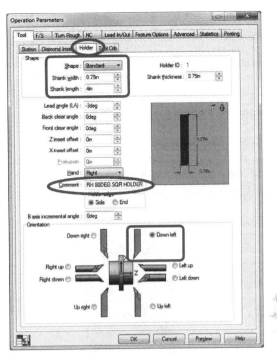

Figure 10.20 The *Holder* tab of the *Tool* tab in the *Operation Parameters* dialog box

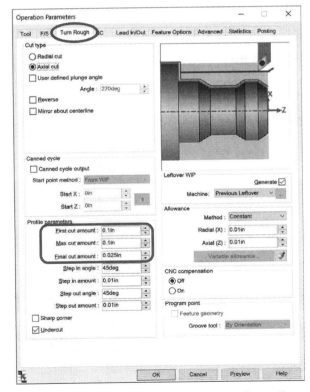

Figure 10.22 The *Turn Rough* tab in the *Operation Parameters* dialog box

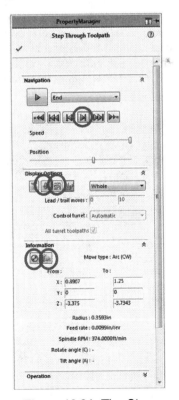

Figure 10.21 The *Step Through Toolpath* dialog box

Step Through Toolpath

Now we take a closer look at the toolpath using the *Step Through Toolpath* capability to better understand the turning operations generated.

We pick two operations for a closer look, *Turn Finish* and *Cut Off*, since both are simple and easier to understand.

In *Turn Finish* operation, the tool moves along the five edges of the OD machinable feature. In *Cut Off* operation, the tool cuts along the straight edge at the rear end face of the part. Reviewing the toolpath will help us understand the G-code to be discussed shortly.

Right click *Turn Finish1* and choose *Step Thru Toolpath*. The *Step Through Toolpath* dialog box appears (Figure 10.21).

Under *Information*, SOLIDWORKS CAM shows the tool movement from the current to the next step in X, Y, and Z coordinates, with the feedrate, spindle speed, and other machining information.

Click the *Step* button ▶ at the center of the *Step Through Toolpath* dialog box (circled in Figure 10.21) to step through the toolpath. You may want to turn on *Show toolpath points* and *Tool Holder Shaded Display* (circled in Figure 10.21) to see the toolpath display similar to that of Figure 10.23.

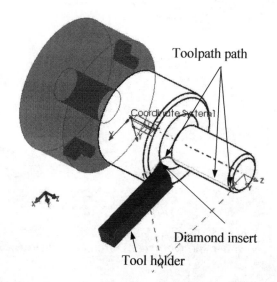

Figure 10.23 Stepping through the toolpath

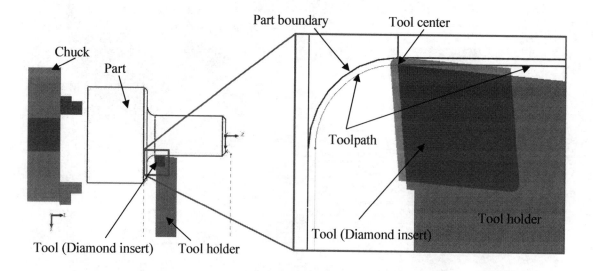

Figure 10.24 Toolpath of *Turn Finish*, a closer look

A closer look at the toolpath of *Turn Finish* (see Figure 10.24) indicates that the toolpath follows the trace of the tool center (i.e., the center of the corner radius of the insert). As a result, in general, the toolpath offsets an amount of tool radius from the part boundary. The XZ coordinates of the six characteristic points of the part boundary, A to F shown in Figure 10.25, referring to the part setup origin are listed in Table 10.1. The toolpath offset by an amount of the tool radius, 0.02in., is especially clear at points B, C, D and E. Note that if you see different results, you may want to check if the radius of the cutter insert has been changed to 0.02in. Note that you may need to click the *Radial or Diameter Coordinate Display* button ◉ and the *Tip or Center Coordinate Display* button 🔧 under Information of the *Step Through Toolpath* dialog box (circled in Figure 10.21) to bring out the XZ coordinates that are consistent with those shown in Table 10.1.

Now we follow the same steps to review the toolpath of the *Cut Off* operation.

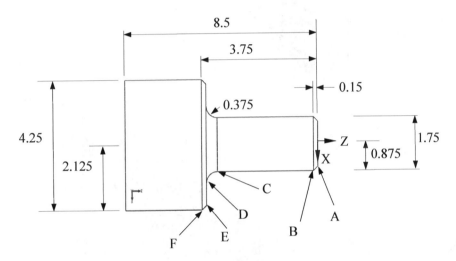

Figure 10.25 The six characteristic points (A to F) of the part boundary

Table 10.1 The XZ Coordinates of the Six Characteristic Points A to F

Point	X and Z Coordinates of Part Boundary	X and Z Coordinates of Toolpath	Offsets in X and Z Coordinates
A	X = 0.725 Z = 0	X = 0.7391 Z = 0.0141	ΔX = 0.0141 ΔZ = 0.0141
B	X = 0.875 Z = −0.15	X = 0.895 Z = −0.15	ΔX = 0.02 ΔZ = 0
C	X = 0.875 Z = −3.375	X = 0.895 Z = −3.375	ΔX = 0.02 ΔZ = 0
D	X = 1.25 Z = −3.75	X = 1.25 Z = −3.73	ΔX = 0 ΔZ = 0.02
E	X = 1.975 Z = −3.75	X = 1.975 Z = −3.73	ΔX = 0 ΔZ = 0.02
F	X = 2.125 Z = −3.9	X = 2.1391 Z = −3.8859	ΔX = 0.0141 ΔZ = 0.0141

Right click *Cut Off1* and choose *Step Thru Toolpath*. Note that the Z coordinates shown in the *Step Through Toolpath* dialog box (Figure 10.26) are −6.559, while the rear end face of the part is 6.5in. to the left of the part setup origin (again, located at the center of the front end face of the part). This is because a groove cutter of 0.118in. in width was selected by SOLIDWORKS CAM for the operation. The right edge of the tool is in contact with the part boundary, as seen in Figure 10.27. Therefore, the toolpath is offset an amount of half cutter width to the left of the part boundary. That is, the Z-coordinates of the toolpath are −6.5−0.118/2 = −6.559, as expected.

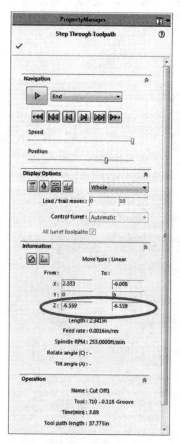

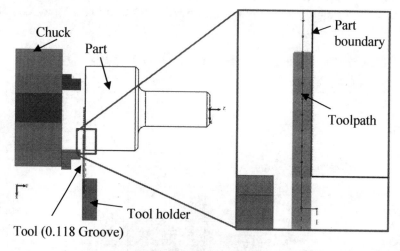

Figure 10.27 Toolpath of *Cut Off*, a closer look

Figure 10.26 The *Step Through Toolpath* dialog box

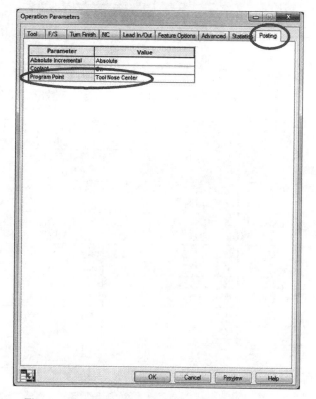

Figure 10.28 The *Posting* tab of the *Operation Parameters* dialog box

10.4 Reviewing the G-code of Turning Sequences

We are now ready to generate and review G-code for *Turn Finish* and *Cut Off* operations. Note that we choose to output the G-code at tool nose center so that the code can be verified with the selected CL data discussed above.

We right click *Turn Finish1* and choose *Edit Definition*. In the *Operation Parameters* dialog box, choose *Posting* tab, and select *Tool Nose Center* for *Program Point* (circled in Figure 10.28). Click *OK* to accept the selection.

Right click *Turn Finish1* and choose *Post Process*. In the *Post Output File* dialog box (Figure 10.29), choose a proper file folder, and enter a file name (for example, *Turn finish.txt*). The *Post Process* dialog box appears (Figure 10.30).

In the *Post Process* dialog box, click the *Play* button ▶ (circled in Figure 10.30) to create G-code (.txt file).

Open the *Turn finish.txt* file from the folder using *Word* or *Word Pad*; see the file contents shown in Figure 10.31(a). It is shown that the NC blocks N8 and N9 moves the cutter to Point A, N11 to Point B, N12 to Point C, N13 to Point D, N14 to Point E, and N16 to Point F. Note that the X locations of the cutter are output as diametral instead of radial determined by the post processor employed.

Repeat the same for the cut off operation. The NC block N8 shows the Z-coordinate of the toolpath; see Figure 10.31(b). Note that the Z-coordinate shown in the G-code is –6.559, which is correct.

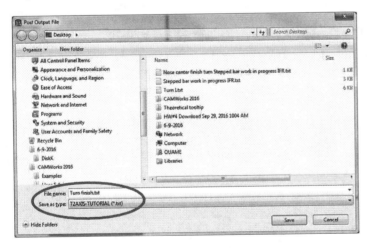

Figure 10.29 The *Post Output File* dialog box

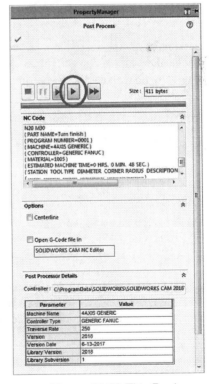

Figure 10.30 The *Post Process* dialog box

We have now completed the lesson. You may save your model for future reference.

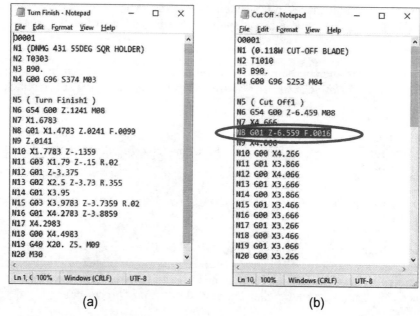

Figure 10.31 Turn G-codes, (a) *Turn Finish* operation, and (b) *Cut Off* operation

10.5 Exercises

Problem 10.1. Repeat the same steps in this lesson to generate turning operations, except that we use the automatic feature recognition capability by clicking the *Extract Machinable Features* button above the graphics area. Is there any redundant machinable feature extracted? Generate toolpath. Are there any noticeable differences in the toolpath of the three turning operations, turn rough, turn finish, and cut off, compared with those discussed in this lesson?

Problem 10.2. Generate NC operations to machine the part shown in Figure 10.32 from a stock of ϕ2.75in.×4.5in. Pick adequate tools (with justifications). Please submit the following for grading:

(a) A summary of the turn operations, including cutting parameters and tools selected.
(b) Screen shots of combined NC toolpaths and material removal simulations.
(c) Is there any material remaining uncut? If so, was it a part design issue or machining issue? What can be done to improve either the part design or machining operations to correct the issue?

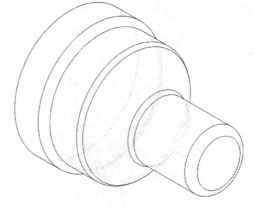

Figure 10.32 The design model of Problem 10.2

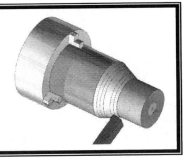

Lesson 11: Turning a Stub Shaft

11.1   Overview of the Lesson

In Lesson 11, we machine a stub shaft, which is similar to that of Lesson 10 but with more turn features, including face, groove, thread, and holes at both ends. Since the two holes are located at the front and rear ends of the part, respectively, we will have to define two turn setups that support turning operations to machine the features from both ends of the bar stock.

Lesson 11 offers a more in-depth discussion in creating turning operations using SOLIDWORKS CAM. You will learn to use automatic feature recognition (AFR) to extract machinable features and make necessary adjustments for a valid turning simulation that can be implemented physically. We will use the same machine as Lesson 10, that is, a lathe of single turret, to machine the stub shaft example shown in Figure 11.1(a) from a bar stock clamped into a three-jar chuck from its rear end; then clamp the front end, to turn machinable features from both ends. Note that only the chuck at the rear end is shown in Figure 11.1(b). The machined part at the end of the material removal simulation of the first turn setup (*Turn Setup1* only) is shown in Figure 11.1(c).

Most of the turning operations generated by SOLIDWORKS CAM are adequate for this example, except for the threading operation. Although we stay with default options and machining parameter values for most of the lesson, we point out a few deficiencies that need to be corrected, including the threading operation. We will take a closer look at the threading operation by stepping through the toolpath and reviewing the G-code to gain a better understanding.

After completing this lesson, you should be able to carry out a turning simulation for applications of most turning features following the same procedures.

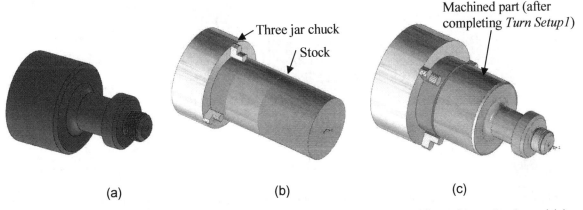

(a)   (b)   (c)

Figure 11.1 The stub shaft example, (a) the design model, (b) bar stock clamped to a chuck, and (c) the material removal simulation (after *Turn Setup1*)

11.2 The Stub Shaft Example

Similar to the stepped bar example, the stub shaft has a bounding cylinder of size ϕ4.25in.×6.5in. In addition to the revolve, chamfer and fillet features, there are holes at the respective two ends; see Figure 11.2(a). Note that the sketch of the revolve feature shown in Figure 11.2(b) is a bit more complex than that of Lesson 10, which leads to additional machinable features, such as groove features. The sketches of the end holes and thread can be seen in Figure 11.2 (c). The thread pitch is 0.125in. (not shown in the figure) with depth 0.451in. Note that the reference features, including a point and a coordinate system, are identical to those of Lesson 10. These features are listed in the feature tree shown in Figure 11.3. Similar to Lesson 10, the coordinate system (*Coordinate System1*) is chosen as the fixture coordinate system. When you open the solid model *Stub shaft.SLDPRT*, you should see the solid features and the reference features listed in the feature tree (see Figure 11.3).

A bar stock of size ϕ4.25in.×8in., made of low carbon alloy steel (1005), as shown in Figure 11.4(a), is chosen for the machining operations. The front end of the stock is extended by 0.25in. from the front face of the part. The rear end of the stock is fixed in the three-jar chuck with an extra length of 1.25in. to be cut off at the end of the turning operations of the first turn setup (similar to that of Lesson 10). Due to the extra material at the front end face, a face turning operation is required.

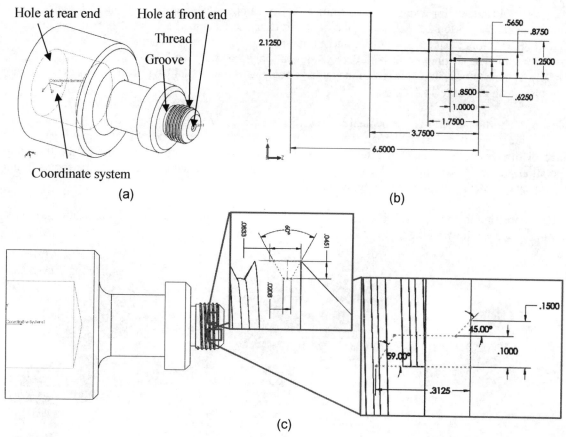

Figure 11.2 The stub shaft solid model, (a) major solid features, (b) sketch and dimensions of the revolve solid feature, and (c) sketches and dimensions of the two end holes and thread

Note that the part setup origins are defined at the center of the front end face of the stock and the origin of the coordinate system, *Coordinate System1*, for the two turn setups, respectively. The first setup cuts all features except for the hole at the rear end, which is machined by the operations included in the second setup.

We first extract machinable features using the automatic feature recognition (AFR) capability.

Most machinable features are extracted, including face, OD, grooves, and ID features (the two holes). The only feature that is not extracted is the thread. We manually create a thread machinable feature.

Overall, there are eight machinable features included in this example, as shown in Figure 11.5. In this lesson, we follow the recommendations of the technology database (TechDB™) for determining machining options, tools, and cutting parameters.

Figure 11.3 Entities listed in the feature tree

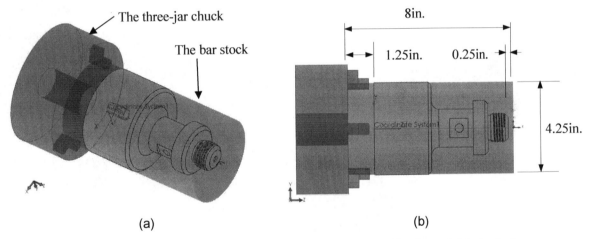

(a)   (b)

Figure 11.4 A bar stock of ϕ4.25in.×8in., (a) stock fixed in a chuck, and (b) stock size

The toolpath of the turn operations for machining the face, OD, and groove features usually consists of rough and finish operations. Hole drilling consists of center drill and drill operations. The rear end hole requires additional operations, such as bore, due to its size.

There are a total of fifteen operations created to machine this stub shaft. Toolpaths of the fifteen operations are shown in Figure 11.6. You may open the example file, *Stub shaft with toolpath.SLDPRT*, to preview the toolpath and turn operations created for this example.

The unit system chosen is IPS.

11.3 Using SOLIDWORKS CAM

Open SOLIDWORKS Part

Open the stub shaft part (filename: *Stub shaft.SLDPRT*) downloaded from the publisher's website. This solid model shown in Figure 11.2 consists of six solid features (revolve, chamfer, fillet, two cut extrudes and a thread), a point, two planes, and a coordinate system (see the solid feature tree in Figure 11.3). As soon as you open the model, you may want to check the unit system and increase the decimals from the default 2 to 4 digits.

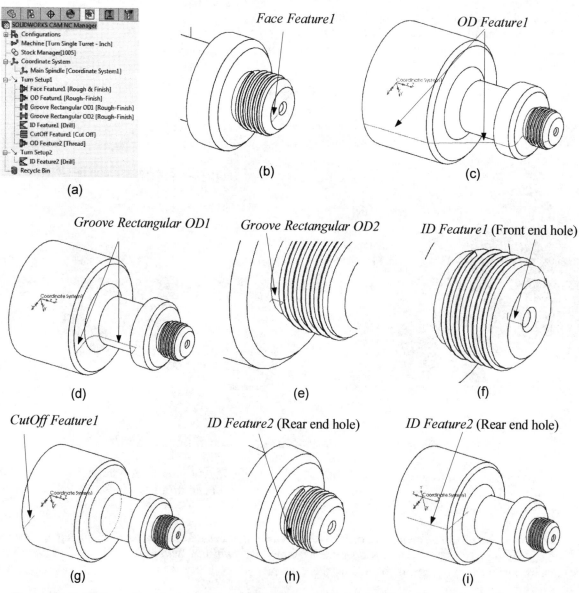

Figure 11.5 The machinable features of the turning operations: (a) SOLIDWORKS CAM feature tree, (b) *Face Feature1*, (c) *OD Feature1*, (d) *Groove Rectangular OD1*, (e) *Groove Rectangular OD2*, (f) *ID Feature1* (front end hole), (g) *CutOff Feature1*, (h) *OD Feature2*, and (i) *ID Feature2* (rear end hole) of *Turn Setup2*

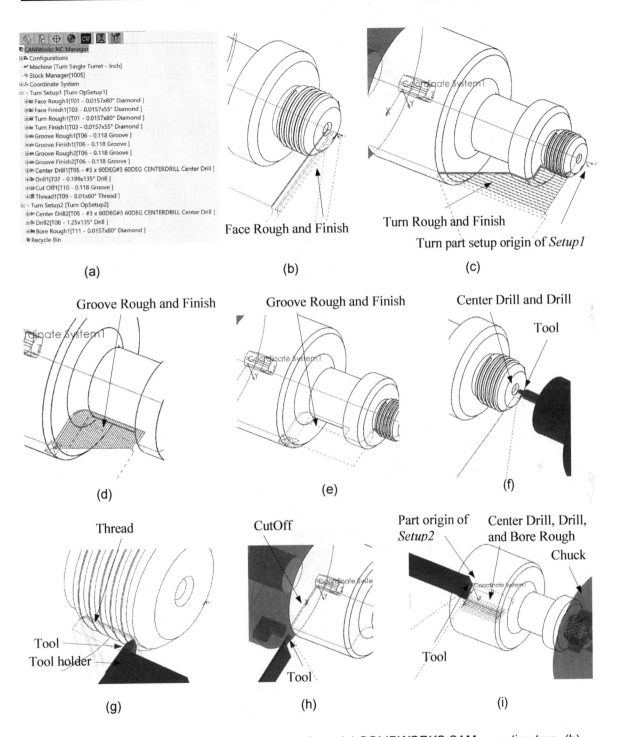

Figure 11.6 The toolpaths of the turning operations, (a) SOLIDWORKS CAM operation tree, (b) *Face Rough1* and *Face Finish1*, (c) *Turn Rough1* and *Turn Finish1*, (d) *Groove Rough1* and *Groove Finish1*, (e) *Groove Rough2* and *Groove Finish2*, (f) *Center Drill1* and *Drill1*, (g) *Thread1*, (h) *Cut Off1*, and (i) *Center Drill2, Drill2,* and *Bore Rough1* of *Turn Setup2*

Select NC Machine

Click the SOLIDWORKS CAM feature tree tab 🔩. Right click *Machine* and select *Edit Definition*. Similar to those of Lesson 10, under *Machine* tab of the *Machine* dialog box, we select *Turn Single Turret - Inch* from the list of *Applicable machines* box, and click *Select*. We choose *Tool Crib2 Rear* under *Available tool cribs* of the *Tool Crib* tab, select *T2AXIS-TUTORIAL* under the *Post Processor* tab, and select *Coordinate System1* for *Main spindle coordinate system* under the *Setup* tab.

Create Stock

From SOLIDWORKS CAM feature tree 🔩, right click *Stock Manager* and choose *Edit Definition*. In the *Stock Manager* dialog box (Figure 11.7), we enter 8in. for the stock length, increasing by 0.25in. to the right of the part from that of Lesson 10. The stock material is *Steel 1005*. We enter the length of the stock outside the part to be –1.25in., as shown in Figure 11.7. As a result, a 0.25in. of extra material appears to the right end face of the part. Accept the stock by clicking the checkmark ✓ at the top left corner. The bar stock should appear in the graphics area similar to that of Figure 11.4.

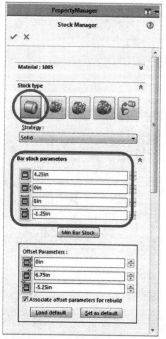

Figure 11.7 The *Stock Manager* dialog box

Turn Setups and Machinable Features

Click the *Extract Machinable Features* button ⬛ above the graphics area. Two setups, *Turn Setup1* and *Turn Setup2*, are created with multiple machinable features extracted.

Under *Turn Setup1*, there are six features extracted: *Face Feature1*, *OD Feature1*, *Groove Rectangular OD1*, *Groove Rectangular OD2*, *ID Feature1* and *CutOff Feature1*. Under *Turn Setup2*, there is one feature extracted, *ID Feature2*. All machinable features are in magenta color, as shown in Figure 11.8. If you click any of these features, you should see them in the graphics area (as lines or curves) like those shown in Figure 11.5. Note that the *Thread1* solid feature was not extracted as a machinable feature. We will add a thread machinable feature manually shortly.

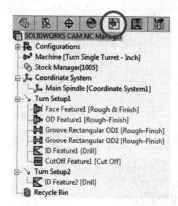

Figure 11.8 The machinable features extracted

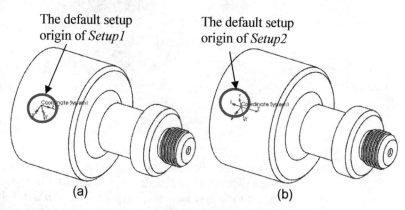

Figure 11.9 The part setup origins of *Setup1* and *Setup2* at the origin of *Coordinate System1*

Click *Turn Setup1* and *Turn Setup2* to locate the respective setup origins in the graphics area. As shown in Figure 11.9, both origins are located at the origin of *Coordinate System1* with Z-axis points in opposite directions. The directions are all good but not the location. We will move the origin of *Turn Setup1* to the front end face of the stock, and keep the origin of *Turn Setup2* as it is. Before we make this change, we will generate operation plans first.

Generate Operation Plan and Toolpath

Click the *Generate Operation Plan* button ⬚ above the graphics area.

Eleven and two operations are generated for *Turn Setup1* and *Turn Setup2*, respectively (see Figure 11.10). All are in magenta color. Expand the operation, for example, *Turn Rough1* to see the associated machinable feature of the operation (in this case, *OD Feature1*). Click *OD Feature1* to show the feature in the graphics area, as shown in Figure 11.10.

Click the *Generate Toolpath* button ⬚ above the graphics area to generate toolpath. Turning toolpaths are generated for all operations.

Relocate Part Setup Origin of *Setup1*

As mentioned earlier, we will move the origin of *Setup1* to the front end face of the stock, which is more practical since this point is easier to access in setting up the stock on a lathe.

Under the SOLIDWORKS CAM operation tree tab ⬚, right click *Turn Setup1* and choose *Edit Definition*. In the *Operation Setup Parameters* dialog box, choose *Origin* tab, and select *Stock Vertex*. The vertex at the right end of the stock in the sketch (circled in Figure 11.11) is selected, which matches our intent. The part setup origin moved to the center of the front end face of the stock—circled in Figure 11.12(a)—as desired.

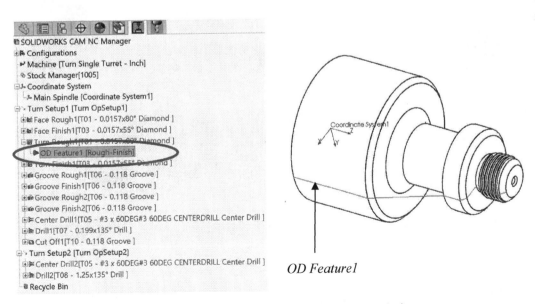

OD Feature1

Figure 11.10 The thirteen operations generated

Click *OK* in the *Operation Setup Parameters* dialog box to accept the change. Click *Yes* to the question in the warning box: *The origin or chuck/fixture location/avoidance parameters has changed, toolpaths need to be recalculated. Regenerate toolpaths now?*

Review the Operations

Next, we review the tool and key machining parameters of a selected operation, *Groove Rough1*.

Under the SOLIDWORKS CAM operation tree tab 🔧, right click *Groove Rough1* and choose *Edit Definition*. In the *Operation Parameters* dialog box, choose *Tool* tab and then *Groove Insert* tab, a groove insert of 0.008in. radius and 0.118in. in width (0.118 Wide GROOVE INSERT) appears (Figure 11.13). Tool holder with groove insert appears in the graphics area (see Figure 11.14). Click the *Holder* tab (see Figure 11.15); a standard holder of shank width 0.75in. and length 4in. has been chosen. Also, the *Down left* is chosen for *Orientation*, which defines the orientation of the cutter suitable to turn the groove feature.

Choose the *Groove Rough* tab of the *Operation Parameters* dialog box. In the *Parameters* area, the *Stepover* is set to *0.05in*, and *Allowance* is set to *0.01in* for both X and Z directions, as shown in Figure 11.16.

The *Stepover* defines the distance of the tool movement along the negative X-direction, similar to the depth of cut in a milling operation. Allowances leave a small amount of material uncut, similar to those of rough mill operations. Click *Cancel* to close the dialog box since we are not making any changes to the *Groove Rough1* operation. You may select other operations to acquire a better understanding of the tools and parameters selected.

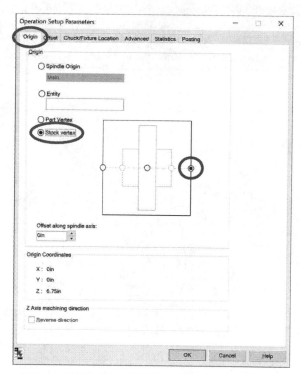

Figure 11.11 The *Origin* tab of the *Tool* tab in the *Operation Setup Parameters* dialog box

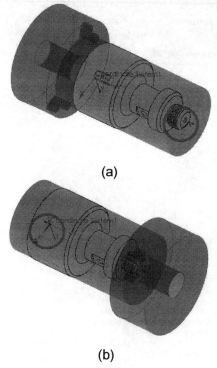

(a)

(b)

Figure 11.12 The part setup origins of (a) *Turn Setup1*, and (2) *Turn Setup2*

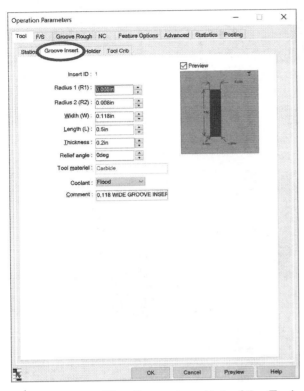

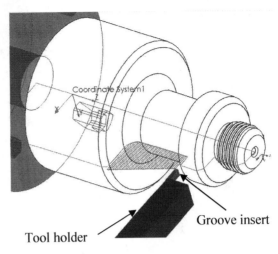

Figure 11.14 Tool holder with groove insert appears in the graphics area

Figure 11.13 The *Groove Insert* tab of the *Tool* tab of the *Operation Parameters* dialog box

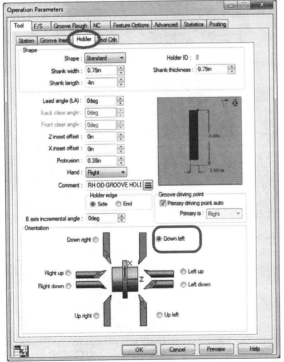

Figure 11.15 The *Holder* tab of the *Tool* tab of the *Operation Parameters* dialog box

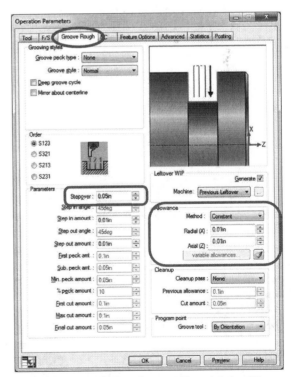

Figure 11.16 The *Groove Rough* tab of the *Operation Parameters* dialog box

Simulate Toolpath

Click the *Simulate Toolpath* button ![Simulate Toolpath icon] above the graphics area to simulate the toolpaths. The material removal simulation appears similar to those of Figure 11.17(a) and Figure 11.17(b) for *Turn Setup1* and *Turn Setup2*, respectively. All look good.

Next, we add a thread operation to *Setup1*, and add a bore operation to *Setup2* for enlarging the hole at the rear end.

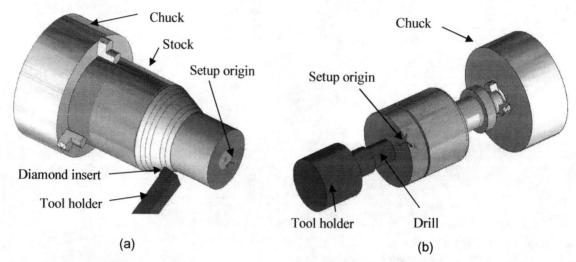

(a)   (b)

Figure 11.17 The material removal simulation, (a) *Turn Setup1*, and (2) *Turn Setup2*

11.4   Cutting the Thread

We now take a look at the thread. Since the thread solid feature was not extracted as a machinable feature, we will have to manually add one.

From SOLIDWORKS CAM feature tree ![icon], right click *Turn Setup1* and choose *Turn Feature* (see Figure 11.18).

In the *New Turn Feature* dialog box (Figure 11.19), choose *OD Feature* for *Type* (should have been chosen by default), select *Thread* for *Strategy*, and then pick the line segments that represent the crest of the thread near the front end of the part in the graphics area (see Figure 11.19). The line segments picked are now listed under *Selected Entities* of the *New Turn Feature* dialog box. And thread profile is created automatically by joining these line segments (joints are shown in the part and listed in the *Selected Entities*). Click the checkmark ✓ to accept the thread feature.

In the SOLIDWORKS CAM feature tree ![icon], an *OD Feature2 [Thread]* is added and is in magenta color under *Turn Setup1*, as shown in Figure 11.20.

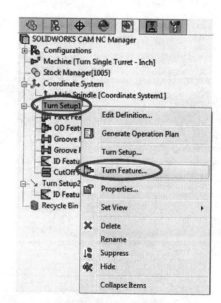

Figure 11.18 Creating a threading machinable feature

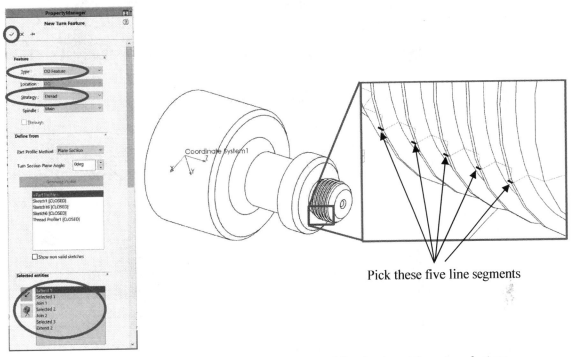

Figure 11.19 Picking the line segments at the front end for a turn feature

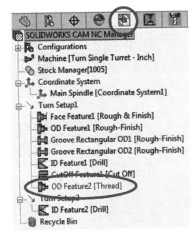

Figure 11.20 *OD Feature2* added to the CAM feature tree

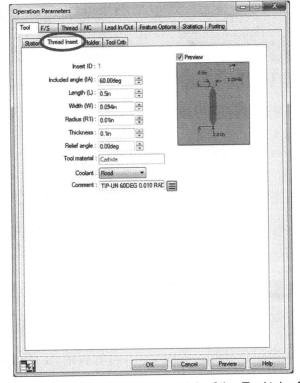

Figure 11.21 The *Thread Insert* tab of the *Tool* tab of the *Operation Parameters* dialog box

Right click *OD Feature2 [Thread]* and choose *Generate Operation Plan*. A *Thread1* operation is now added to the operation tree under *Turn Setup1*, again in magenta color.

Right click *Thread1* and choose *Generate Toolpath*. Toolpath will be generated similar to that of Figure 11.6(g).

Next, we take a closer look at the thread operation and thereafter, the toolpath.

Under the SOLIDWORKS CAM operation tree tab ▇ , right click *Thread1* and choose *Edit Definition*. In the *Operation Parameters* dialog box, choose *Tool* tab and then *Thread Insert* tab, a thread insert of 0.01×60° (radius 0.01in. included angle 60 degrees, thickness 0.1in.) appears (Figure 11.21).

Click the *Holder* tab (see Figure 11.23); a standard holder of shank width 0.75in. and length 4in. has been chosen. Also, the *Down left* is chosen for *Orientation*, which defines the orientation of the cutter suitable to turn the thread feature. Tool holder with thread insert appears in the graphics area (see Figure 11.22).

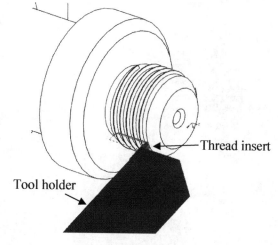

Figure 11.22 Tool holder with thread insert appears in the graphics area

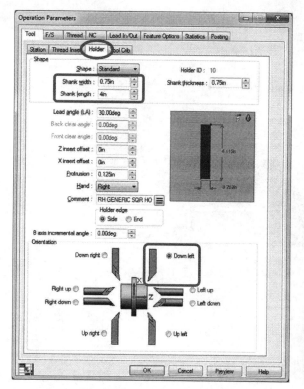

Figure 11.23 The *Holder* tab of the *Tool* tab of the *Operation Parameters* dialog box

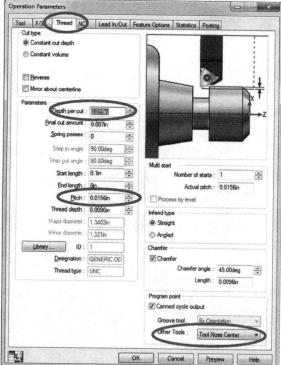

Figure 11.24 The *Thread* tab of the *Operation Parameters* dialog box

Click the *Thread* tab (Figure 11.24), and note that the *Depth per cut* is *0.015in.*, *Pitch* is *0.0156in.* (which are not consistent with the dimensions of the thread solid feature and will be corrected next), and *Tool Nose Center* is chosen for *Program point*. The other option for *Program point* is *Tool Nose Tip*. Different options affect the G-code, which will be discussed later in this lesson.

Click the *Feature Options* tab, and click *Parameters* (see Figure 11.25). In the *OD Profile Parameters* dialog box (Figure 11.26), enter *0.0451in.* for *Thread depth* and *0.125in.* for *Pitch* so as to make the thread size consistent with that of the solid feature; see dimensions of the thread in Figure 11.2(c).

Note that *Maximum dia (D1): 1.3402in* and *Minimum dia. (D2): 1.25in* circled in Figure 11.26 are the major and minor diameters of the thread, respectively, extracted by SOLIDWORKS CAM.

Click *OK* in the *OD Profile Parameters* dialog box. The *Pitch* is now *0.125in.* under the *Thread* tab of the *Operation Parameters* dialog box. Click *OK* in the *Operation Parameters* dialog box to accept the changes.

Step Through Toolpath

Now we take a closer look at the toolpath to better understand the toolpath (and later the G-code) of the threading operation.

Right click *Thread1* and choose *Step Thru Toolpath*. The *Step Through Toolpath* dialog box appears (Figure 11.27).

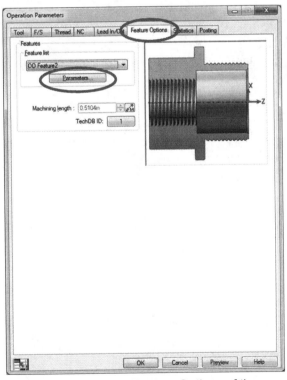

Figure 11.25 The *Feature Options* of the *Operation Parameters* dialog box

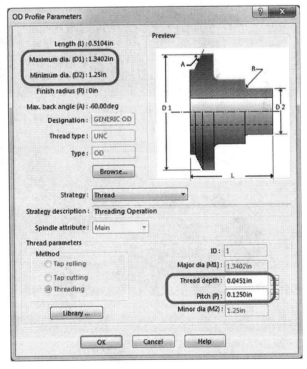

Figure 11.26 The *OD Profile Parameters* dialog box

Under *Information*, SOLIDWORKS CAM shows the tool movement from the current to the next step, the feedrate, and spindle speed, among others.

Click the *Step* button 🔳 at the center (circled in Figure 11.27) to step through the toolpath. You may want to turn on *Show toolpath points* and *Tool Holder Shaded Display* to see the toolpath display similar to that of Figure 11.28.

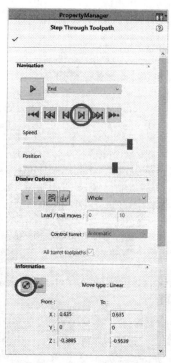

A zoom in view of the toolpath (see Figure 11.28) indicates that the toolpath follows the trace of the tool center (center of the tool nose radius of the insert). As a result, in general, the toolpath offsets an amount of tool nose radius from the part boundary.

For example, the XZ coordinates of the last cutting pass, as shown in Figure 11.29(a) are from (0.635, −0.3885) to (0.635, −0.9539). You may click the *Radial or Diameter Coordinate Display* button ⊘ (circled in Figure 11.27) to toggle the numeric values displayed for the X-coordinate between radius and diameter.

The X coordinates indicate that the last pass is 0.635in. away from the part set up origin, as shown in Figure 11.29(a), which is the minor radius of the thread plus the tool radius; i.e., 0.625+0.01 = 0.635, as it should be.

Recall that we chose *Tool Nose Center* for *Program point*; hence, XYZ coordinates of the toolpath are generated at the tool nose center. If you choose *Tool Nose Tip* for *Program point*, the X coordinate of the last cutting pass should be 0.625, which is simply the minor radius of the thread.

Figure 11.27 The *Step Through Toolpath* dialog box

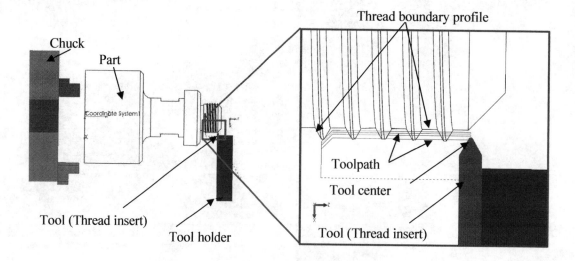

Figure 11.28 Toolpath of *Thread1*, a closer look

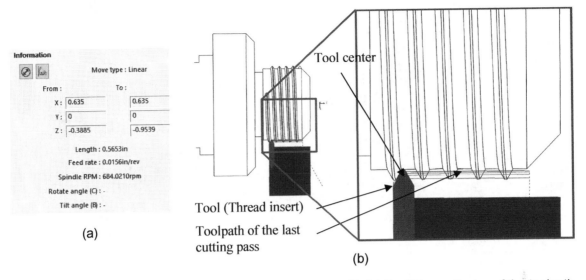

Tool center

Tool (Thread insert)

Toolpath of the last
cutting pass

(a)

(b)

Figure 11.29 The last cutting pass of the toolpath of *Thread1*, (a) the XZ coordinates of the toolpath,
and (b) a zoom in view of the cutter location

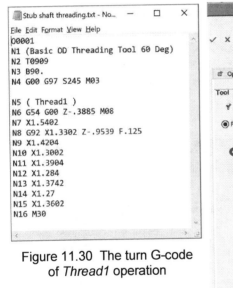

Figure 11.30 The turn G-code
of *Thread1* operation

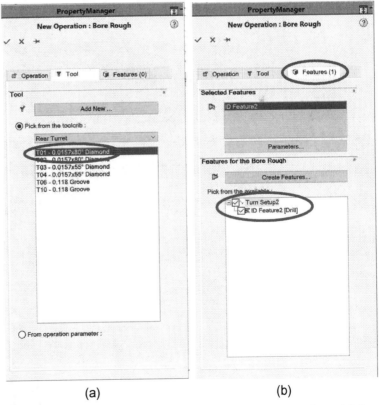

(a)   (b)

Figure 11.31 The *New Operation: Bore Rough* dialog box, (a) the
Tool tab, and (b) the *Features* tab

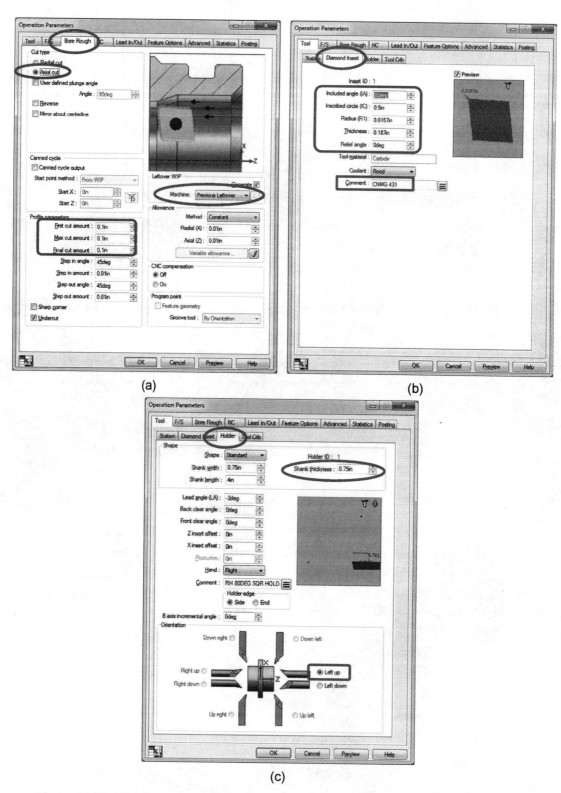

(a)

(b)

(c)

Figure 11.32 The *Operation Parameters* dialog box, (a) the *Bore Rough* tab, (b) the
Diamond Insert tab, and (c) the *Holder* tab

G-code of the Threading Operation

Right click *Thread1* and choose *Post Process* to generate the G-code. Enter, for example, *Stub shaft thread* for filename. Open the *Stub shaft threading.txt* file (see file contents shown in Figure 11.30). It is shown that the NC block N14 moves the cutter along the last cutting pass, in which the X-coordinate is 1.27 (= 2×0.635), which is correctly converted. Again, the X locations of the cutter are output as diametral instead of radial determined by the post processor employed.

11.5 Boring the Hole

We now add a boring operation to the hole at the rear end under *Turn Setup2*. Under the SOLIDWORKS CAM operation tree tab 🔧, right click *Turn Setup2* and choose *Turn Bore Operations* and then *Bore Rough*. In the *New Operation: Bore Rough* dialog box, choose *T01 – 0.0157×80° Diamond* for tool, and click the *Features* tab to choose *ID Feature2 [Drill]*. Click the checkmark ✔ to accept the operation.

Click *OK* to the SOLIDWORKS CAM message: *The current insert/holder orientation is not allowed for the Main Spindle while using the Rear Turret. SOLIDWORKS CAM has selected a valid orientation. Please review this selection on the Holder page and change it if required.*

This is because the same tool (T01) was used for operations of *Turn Setup1*, in which *Down left* has been chosen for orientation. The down left orientation is not suitable for the hole boring operation and must be changed.

In the *Operation Parameters* dialog with the *Bore Rough* tab selected, we stay with the default values and selections (for example, *Axial cut*; and *0.1in.* for *First*, *Max*, and *Final cut amount*). Also, we choose *Previous Leftover* for *Machine*—see Figure 11.32(a), which is left over from the *Drill2* operation.

Click the *Tool* tab, and then, *Diamond Insert* tab; see Figure 11.32(b). We use the insert currently selected (*CNMG 431*).

Click the *Holder* tab; see Figure 11.32(c). We reduce the shank thickness to *0.25in.* to avoid collision. Note that *Left up* is chosen for *Orientation*, which is what the earlier message was referring to.

Click *OK* to accept the operation, and click *Add* to the warning message: *Tool parameters have changed. Select Add to create a new tool. Select Change to modify the tool for all operations sharing this tool.* A new operation, *Bore Rough1*, is now listed in the operation tree.

Under the SOLIDWORKS CAM operation tree tab 🔧, right click *Bore Rough1* and choose *Generate Toolpath*. A bore rough tool path appears in the graphics area, similar to that of Figure 11.6(i). Choose *Simulate Toolpath* to carry out a material removal simulation, especially, for *Turn Setup2* like that of Figure 11.33.

We have completed the exercise. You may save your model for future references. Close the model.

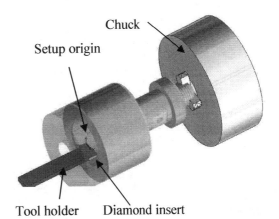

Figure 11.33 The material removal simulation of *Turn Setup2*

11.6 Exercises

Problem 11.1. Generate NC operations to machine the part shown in Figure 11.34 from a stock of ϕ4in.×7.5in. Pick adequate tools (with justifications). Please submit the following for grading:

(a)   A summary of the turn operations, including cutting parameters and tools selected.
(b)   Screen shots of combined NC toolpaths and material removal simulations.
(c)   Is there any material remaining uncut? If so, was it part design issue or machining issue? What can be done to improve either the part design or machining operations to correct the issue?

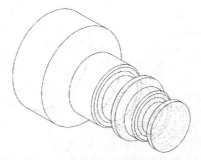

Figure 11.34 The design model of Problem 11.1

Problem 11.2. Repeat Problem 11.1 for the part shown in Figure 11.35 from a stock of ϕ4in.×6in.

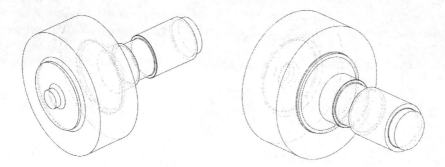

Figure 11.35 The design model of Problem 11.2

Problem 11.3. Repeat Problem 11.1 for the part shown in Figure 11.36 from a stock of ϕ10.5in.×16in.

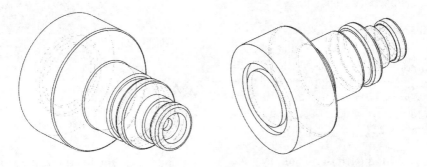

Figure 11.36 The design model of Problem 11.3

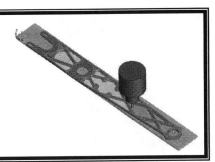

Lesson 12: Machining a Robotic Forearm Member

12.1　Overview of the Lesson

So far, we have discussed machining simulation using SOLIDWORKS CAM. All have been carried out in a virtual environment. All seem to be working really well. The biggest question is: will the machining simulation generated in computers work swiftly to cut the parts on the shop floor as desired? How do we take the NC programs post processed by using SOLIDWORKS CAM to CNC machines and physically cut parts? Will the G-code work without a glitch? Will the cutters follow the toolpath without running into fixtures? Will the machining operations yield a desired physical part that meets the requirements? What do we have to do and what do we have to consider before we transport the results of virtual machining to CNC machines that physically cut the part? The bottom line question is: How do we go from virtual to physical machining?

Going from virtual machining to physical machining is no small matter. Machining simulation that works perfectly in the virtual environment may not work physically without careful planning. Many issues need to be taken into consideration. Some of the major factors include choosing adequate fixtures to firmly hold the stock to the jig table, checking to ensure that cutters would not collide with fixtures and jig table, ensuring cutters and machining parameters are adequate, verifying that the syntax and semantics of the G-code are compatible with the CNC machine at the shop floor, among others.

Although this book is not written to address the practical aspects of CNC machining, having practical experience or knowing the basic factors in physically cutting parts using CNC machines enhances our ability to create viable machining operations in a virtual environment that hold a higher probability of success.

In this lesson (Lesson 12) and Lesson 13, we discuss the transition from virtual to physical machining.

Although we do not intend to provide a comprehensive discussion on the topic, in these two lessons we point out a few important factors that may be beneficial to you when you are taking the G-code to a CNC machine to cut parts physically. In the current lesson, Lesson 12, we present a milling case study, which involves machining a robotic forearm member on a HAAS 3-axis mill using NC part programs generated in SOLIDWORKS CAM. In Lesson 13, we focus on turning, in which we generate turning operations using SOLIDWORKS CAM for cutting a scaled baseball bat by turning a cylindrical stock on a HAAS CNC lathe. Both cases are extracted from student projects.

Since we have learned to use SOLIDWORKS CAM to generate virtual machining simulation for milling and turning operations, we will not repeat such details in Lessons 12 and 13. Instead, we simply present the machining operations created, point out a few adjustments that may be of interest to your own applications, and focus more on the practical aspects of the machining operations.

After completing this lesson, you should acquire basic knowledge in transition from virtual machining simulation to physical milling operations. In addition to pointing out key factors for your consideration, we state a few lessons learned that would help you avoid similar pitfalls. Although we employed HAAS CNC mill and lathe for the two lessons, and some of you may not even have access to a CNC machine, we hope the information offered in these two lessons brings your virtual machining skill to the next level by taking practical aspects into consideration when you create machining simulation in virtual environment.

12.2 The Forearm Members of the SoRo Robotic Arm

The part to be machined is a forearm member of the 2019 SoRo Robotic Arm, which is an integrable component of Sooner Rover (SoRo). Sooner Rover is a student designed and fabricated rover, with which OU engineering students entered the University Rover Challenge (URC, urc.marssociety.org) hosted by the Mars Society (www.marssociety.org). The URC is a challenging competition that requires the rover with both finer manipulation and much greater lifting capacity. In the first two years at the URC, SoRo has struggled to produce a robotic arm that is well suited to the competition requirements. The 2017 URC arm—see Figure 12.1(a)—suffered from fabrication and execution deficiencies, and the 2018 URC arm—see Figure 12.1(b)—had a severely limited range of motion which necessitated a stowed position with an undesirably high center of gravity.

To prepare for the 2019 URC competition, the SoRo significantly redesigned the structural forearm members—see Figure 12.2(a)—firstly to accommodate jack-screw tensioning mechanisms, and secondly to save weight through lightening holes. The forearm members are exact mirror copies and symmetrical about the longitudinal axis, so two identical parts can be made. A solid model of the forearm member, which shows the L and T shaped profiles of the trusses that provide stability in both the lateral and transverse dimensions, is shown in Figure 12.2(b).

The solid model of the forearm members, as shown in Figure 12.2(b), consists of irregular pockets, slots, and holes. The bounding box of the forearm members is 22.34in.×2.94in.×0.25in. Aluminum 6061-T6 is assumed to reduce weight.

(a)   (b)

Figure 12.1 The SoRo Robotic Arms competed in the University Rover Challenge, (a) 2017 URC arm, and (b) 2018 URC Arm

Although there are many machinable features, as illustrated in the section views of the part—see Figure 12.2(c)—they are all 2.5-axis mill features that can be recognized automatically by using the AFR capability of SOLIDWORKS CAM. Most machining operations automatically generated by SOLIDWORKS CAM are applicable. However, some changes are necessary to support an efficient and viable physical machining on the shop floor.

Note that the redesigned forearm members have a 25% mass savings compared to the 2018 design. Structural integrity of the arm member is verified for the maximum operational loading condition applied to this part using finite element analysis capability offered by SOLIDWORKS Simulation. A safety factor of 5.57 has been achieved. The SoRo team decided to create machining simulation and physically produce two forearm members.

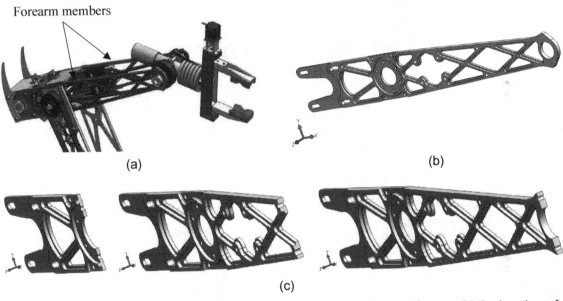

Figure 12.2 The forearm members, (a) located on both sides of the robotic arm, (b) the iso view of the solid model, and (c) three section views showing the major machinable features

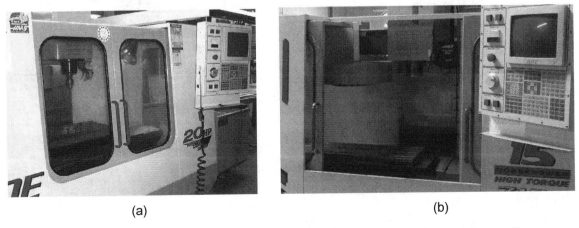

Figure 12.3 The HAAS Mill VF-0E, (a) exterior view, and (b) inside view of the mill

12.3 Manufacturing of the Rover Forearm Members

A HAAS VF-0E mill shown in Figure 12.3(a) is employed to machine the part. The VF series is a 3-axis mill (with an optional rotary table that offers a 4th axis tool motion) with a work envelope 30in.×16in.×20in., which is large enough to accommodate the physical size of the forearm members.

The forearm member will be cut from a rectangular stock of 6061-T6. Since the part is narrow and thin, and requires contour profile milling operations for its outer boundary, a simple setup using standard fixtures, such as vise, clamp, or end block, are not able to firmly clamp the stock to the jig table for the entire machining operation, in particular, cutting the profile boundary of the part. A custom fixture is required, and the overall machining strategy must be carefully determined to completely eliminate chatter concerns during all significant material removal steps. Developing a viable machining strategy and determining use of the adequate fixtures are among the most important aspects to consider while transitioning from virtual to physical machining.

A strategy for machining the forearm members is outlined in the following steps.

Step 1. Place the rectangular stock of 25in.×3in.×0.375in. in aligned double vises mounted on the jig table. Double vises are necessary due to length of part.
Step 2. Face mill one side of the stock for 0.125in.
Step 3. Drill all 12 through-holes and cut all four slots while the stock is in vises. The stock is now similar to that of Figure 12.4(a).
Step 4. Design and machine a custom fixture shown in Figure 12.4(b) to which the stock-in-progress can be attached. The fixture has threaded holes corresponding to screw mounting holes in part design.
Step 5. Bolt stock-in-progress to fixture.
Step 6. Machine part boundary profile.
Step 7. Machine interior pockets.
Step 8. Machine the large hole at the right end.

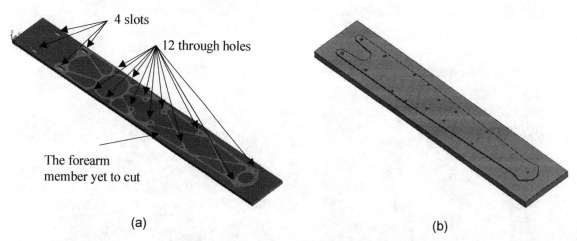

4 slots

12 through holes

The forearm member yet to cut

(a)   (b)

Figure 12.4 Illustration of machining strategy, (a) the stock-in-progress with slots cut and through holes drilled, and (b) the solid model of the custom fixture

12.4 Five Machining Setups

Following the machining strategy stated above, five setups are in place. They are Setup A that machines the rectangular stock to be stock-in-progress by carrying out face milling, slot cutting, and hole drilling;

i.e., Steps 2 and 3. Setup B machines the fixture; i.e., Step 4. Setup C cuts part boundary by carrying out milling operations including contour profile milling; i.e., Step 6. Setup D machines interior pockets (Step 7). And the last setup, Setup E, machines the large hole at the right end (Step 8).

Virtual machining operations, including Setups A, C, D, and E, can be found in the part file, *Forearm Member with Toolpath.SLDPRT*. In addition, machining operations generated for Setup B can be seen in the part file, *Forearm Fixture with Toolpath.SLDPRT*.

Setup A

Seven machining operations are created under Setup A, as listed in Figure 12.5(a). Face Mill1 removes 0.125in. thick material on the top face of the raw stock using a 0.75in. end mill. Rough Mill1/Contour Mill1 and Rough Mill2/Contour Mill2 cut the two slot pairs, respectively. Drill1 and Drill2 drill total eighteen holes of diameters 0.2in.

The material removal simulation of the seven operations under Setup A is shown in Figure 12.5(b). The G-code was uploaded to the HAAS mill to carry out the machining operations on a raw stock of 25in.×3in.×0.375in. clamped in aligned double-vises, as shown in Figure 12.5(c).

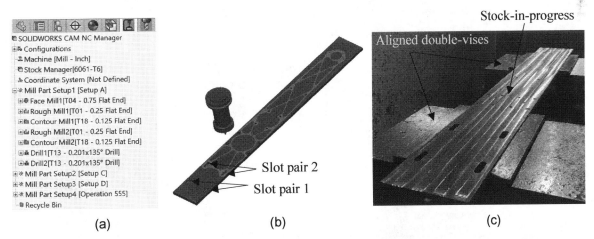

(a)   (b)   (c)

Figure 12.5 Setup A, (a) machining operations generated in SOLIDWORKS CAM, (b) material removal simulation, and (c) the stock-in-progress physically machined

Setup B

Six machining operations are created under Setup B—see Figure 12.6(a)—in which the forearm fixture is cut from a 6061-T6 rectangular stock of 25in.×5in.×0.75in. Rough Mill1 and Contour Mill1 removes a layer of 0.125in. material outside the boundary profile of the part, as shown in Figure 12.6(b). Center Drill1 and Drill1 drill eleven threaded holes of diameters 0.2in., and Center Drill2 and Drill2 drill four threaded holes of diameters 0.25in. to which the workpiece can be attached. The fixture has threaded holes corresponding to screw mounting holes in part design.

The material removal simulation of the six machining operations under Setup B is shown in Figure 12.6(b). The G-code was uploaded to the HAAS mill to carry out the machining operations on a raw stock of 25in.×5in.×0. 75in. clamped in aligned double vises, as shown in Figure 12.6(c).

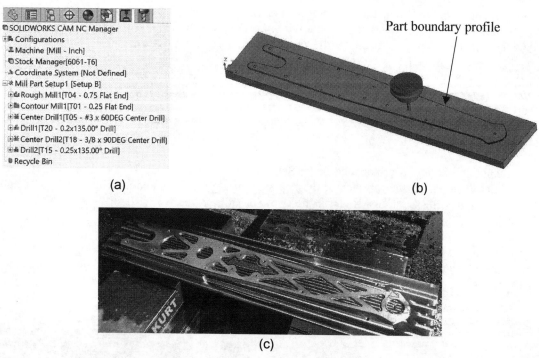

(a)   (b)

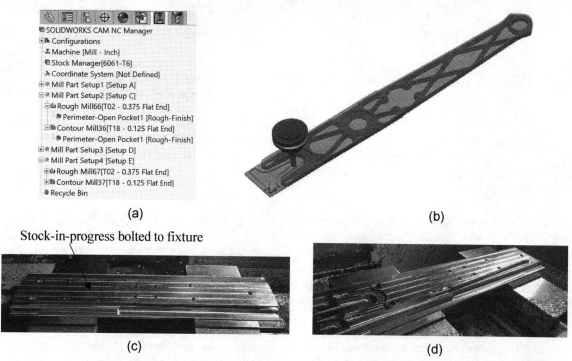

(c)

Figure 12.6 Setup B, (a) machining operations generated in SOLIDWORKS CAM, (b) material removal simulation, and (c) the forearm fixture physically machined

(a)   (b)

Stock-in-progress bolted to fixture

(c)   (d)

Figure 12.7 Setup C, (a) machining operations generated in SOLIDWORKS CAM, (b) material removal simulation, (c) stock-in-progress bolted to fixture, and (d) part profile boundary machined

Setup C

Two machining operations are created under Setup C—see Figure 12.7(a)—in which the boundary profile of the part is cut from the stock-in-progress obtained after completing operations defined in Setup A.

Rough Mill66 uses a 3/8in. end mill to remove a layer of 0.25in. material of a perimeter-open pocket, which covers the area between the boundary profile of the part and the boundary of the stock-in-progress, as shown in Figure 12.7(b). Contour Mill36 employs a 1/8in. end mill to move around the part profile boundary for a finish cut.

At the CNC mill, the stock-in-progress is bolted into the custom fixture obtained from Setup B, as shown in Figure 12.7(c). The G-code of Setup C was uploaded to the HAAS mill to carry out the two machining operations, Rough Mill66 and Contour Mill36, on the stock-in-progress. The physically machined stock with part boundary profile clearly visible is shown in Figure 12.7(d).

Setup D

Setup D cuts 24 pockets, 3 slots, one counterbore hole, and six holes, as shown in Figure 12.8.

The machining strategy for pocket and slot machinable features defined in the TechDB™ is Rough-Rough(Rest)-Finish, resulting in three machining operations, two rough mills and one contour mill. As an example, Rough Mill3, Rough Mill4, and Contour Mill3 are created for machining *Irregular Pocket1* (next to the counterbore hole), as seen in Figure 12.9(a). The first rough mill is a volume milling operation that removes most material of the pocket, for example, the toolpath of Rough Mill3 that cuts the pocket (*Irregular Pocket1*) is shown in Figure 12.9(b). The second rough mill is a rest mill (or local milling) that cleans up the corner area of the pocket; see toolpath of Figure 12.9(c). The contour mill performs profile milling that goes around the boundary profile of the pocket to complete the finish cut, as shown in Figure 12.9(d).

As a result, there are 54 rough mill and 27 contour mill operations that cut the respective 24 pockets and 3 slots, in addition to two contour mill operations that machine the counterbore hole, and one drill operation that drills the six holes (pointed out in Figure 12.8). A partial list of the operations under Setup D is shown in Figure 12.10(a).

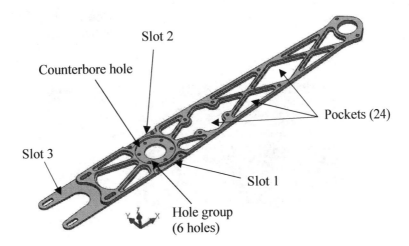

Figure 12.8 Three slots, 24 pockets, one countersink, and six holes to be machined under Setup D

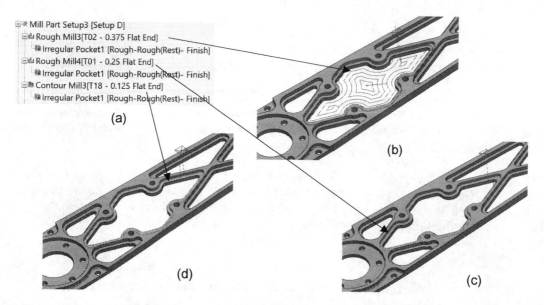

(a)

(b)

(d)

(c)

Figure 12.9 Toolpaths of machining operations cutting *Irregular Pocket1*, (a) toolpath of Rough Mill3 (volume milling), (b) toolpath of Rough Mill4 (local milling), and (c) toolpath of Contour Mill3 (finish cut)

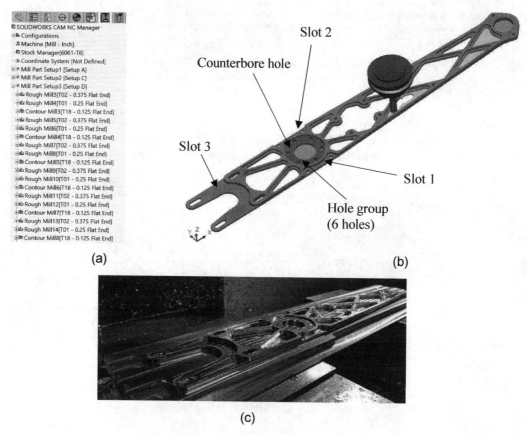

(a)

(b)

(c)

Figure 12.10 Setup D, (a) machining operations generated in SOLIDWORKS CAM, (b) material removal simulation, and (c) part physically machined after completing operations of Setup D

The material removal simulation of the machining operations under Setup D is shown in Figure 12.10(b). The G-code was uploaded to the HAAS mill to carry out the machining operations. The physically machined part after Setup D is shown in Figure 12.10(c).

Note that cutters chosen by TechDB™ for the pocket and slot milling operations are determined mainly by the size of the features among other factors. Different size cutters were chosen by TechDB™, which increases the frequency of tool changes. Tool changes take time. Excessive tool change is in general less desirable.

To minimize the frequency of tool change, we unify the use of cutters, in which 3/8in., 1/4in., and 1/8in. end mills are employed for the two rough mill and contour mill operations, respectively. The order of the operations is adjusted so that rough mill operation of volume milling that employs the same 3/8in. end mill for all pockets and slots are in one group. Similar arrangement is made for all rough mill of local milling using the 1/4in. mill. The same is done for the contour mill operations using the 1/8in. cutter. G-code was generated after the operations were properly grouped to minimize the frequency of tool exchange. As a result, the NC word M06 for tool change is minimized to eliminate tool changes.

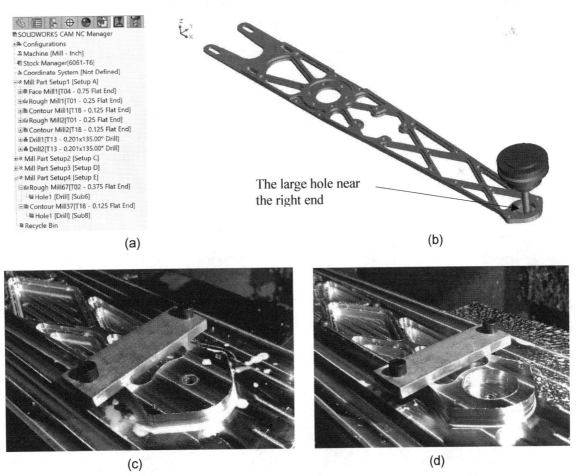

(a)

(b)

The large hole near the right end

(c)

(d)

Figure 12.11 Setup E, (a) machining operations generated in SOLIDWORKS CAM, (b) material removal simulation, (c) bracket bolted to the fixture near the hole, and (d) hole machined

Setup E

Setup E cuts the large hole near the right end of the part shown in Figure 12.11(b). Two mill operations, Rough Mill67 and Contour Mill37—see Figure 12.11(a)—are created for the setup. The material removal simulation of the two machining operations is shown in Figure 12.11(b). A small bracket is bolted to the fixture near the hole to hold the stock firmly to the fixture; see Figure 12.11(c). This setup is necessary to minimize chatter. The completely machined part is shown in Figure 12.11(d).

12.5 Modifications of SOLIDWORKS CAM Operations

As should be expected, the machining operations automatically generated by SOLIDWORKS CAM will not always work without modifications. In this section, we point out major changes made in the milling operations initially generated by SOLIDWORKS CAM for machining the forearm member.

There are 24 irregular pockets, 4 irregular slots, 2 obround pockets, 1 hole, 4 hole groups, and 1 counterbore hole machinable features automatically extracted by SOLIDWORKS CAM, as listed under SOLIDWORKS CAM feature tree; see Figure 12.12(a). Milling operations are automatically generated— part of the operations are shown in Figure 12.12(b)—that machine the part to be like that of Figure 12.12(c), which does not completely support the machining strategy discussed before. Moreover, some areas, including the material outside the profile boundary of the part, were not removed; see Figure 12.12(c). The machinable features and mill operations can be found in the part file: *Forearm Member SOLIDWORKS CAM Toolpath.SLDPRT*.

The following changes have been made to the machining operations of the forearm member to support the machining strategy discussed above.

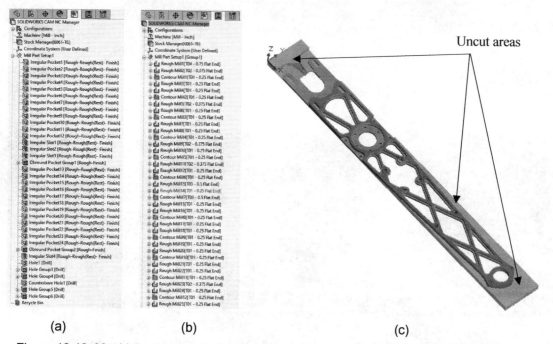

      (a)   (b)   (c)

Figure 12.12 Machining part setup determined by SOLIDWORKS CAM automatically, (a) list of machinable features under SOLIDWORKS CAM feature tree, (b) part of machining operations under SOLIDWORKS CAM operation tree, and (c) a complete material removal simulation

First, a sketch of the same size as the raw stock was added, which is necessary to define as a face machinable feature shown in Figure 12.13(a) for generating a face milling operation under Setup A.

Second, as shown in Figure 12.12(c), no toolpath was generated for some areas of the part. To alleviate the problem, we manually added a perimeter-open pocket, which covered the area between the perimeter of the forearm member and the outside boundary of the raw stock, as shown in Figure 12.13(b). The added perimeter-open pocket covers the uncut areas shown in Figure 12.12(c). Note that *Irregular Slot4*, which overlaps partially with the perimeter-open pocket, was deleted; see Figure 12.13(c).

The third change was creating additional setups and renaming the setups as Setups A to E and assigning machinable features to their respective setups. Thereafter, we generate toolpath using capability of SOLIDWORKS CAM.

Finally, to minimize tool changes, we assigned cutters of 3/8in., 1/4in., and 1/8in. for the two rough mills, and contour mill, respectively, that machine 24 pockets and 3 slots of the forearm member in Setup D, as mentioned earlier. Some of the problematic operations in magenta color—mostly rough-rest, as shown in Figure 12.12(b)—were fixed by changing the cutter to 1/4in. end mill.

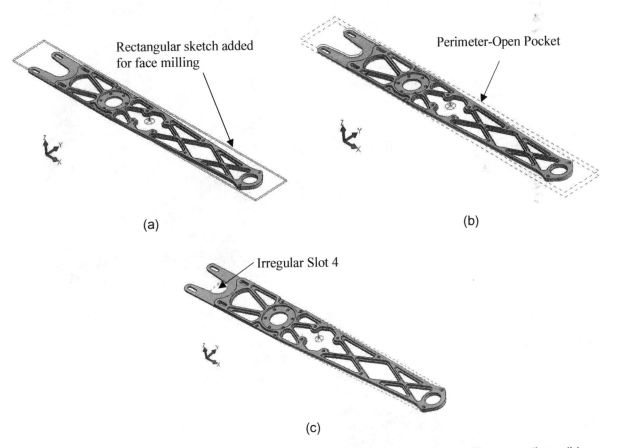

Figure 12.13 Modifying machinable features, (a) adding a sketch for face milling operations, (b) adding a perimeter-open pocket feature, and (c) deleting three irregular slots

12.6 Post-Processing and G-code Modification

After making the adjustment and changes, we convert the toolpaths to G-code using a post process, MILL/HAAS_VF3, which came with SOLIDWORKS CAM, as shown in Figure 12.14. Since a HAAS post processor is employed, the G-code converted was mostly compatible with HAAS mill. Only minimum change for syntax was necessary, such as adding a % sign at the beginning and end of the NC program.

12.7 Machined Part Clean-up

The parts machined usually have rough edges, which pose a real risk of cutting the worker that handles them. One simple solution to this problem is a simple deburring tool, as shown in Figure 12.15. By using the deburring tool, all the edges were cleaned up manually. This makes the part safer for handling. Another possible solution to this would have been to incorporate a chamfering operation in the CNC process to take down all the edges prior to removing it from the machine.

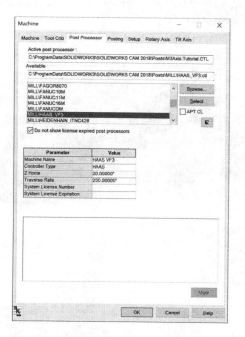

Figure 12.14 Selecting HAAS_VF3 for post processor

Figure 12.15 A simple deburring tool

12.8 Lessons Learned

SOLIDWORKS CAM is nothing but a computer software, a really good one. It is like any tools that we learned to use to get our job done right with better quality and less time. Knowing which buttons to click and which options to choose in SOLIDWORKS CAM is important. It is much more important to know how to maximize the use of the software to our advantage. By transferring from virtual to physical environment discussed in this lesson, you have seen there are several important factors to take into consideration. More specifically, here are lessons learned from machining the forearm members.

(1)   It is desirable for users of CAM tools, such as SOLIDWORKS CAM, to acquire practical experience of using CNC machines to physically cutting parts. This practical experience should

guide and assist CAM software users to generate machining operations that are physically viable and have higher probability in cutting parts successfully.

(2)   It is indispensible for CAM software users to develop a machining strategy and create virtual machining operations that support the implementation of the machining strategy at the shop floor.

(3)   Generating machining operations to cut the design part may not be sufficient. In many situations, custom fixtures are required. Designing and machining fixtures are important steps of a successful part machining.

(4)   Machining operations generated from SOLIDWORKS CAM may not be suitable and may not support the machining strategy developed. Machining operations and associated toolpaths automatically generated by SOLIDWORKS CAM (and any other CAM software tools) must be examined carefully to verify if they are viable physically. Usually, changes are made to support an accurate and efficient machining assignment that supports the machining strategy developed.

(5)   In many situations, G-code converted from a chosen post processor of SOLIDWORKS CAM must be verified carefully to ensure the syntax and semantics are compatible with the CNC machine.

We have completed this lesson. We hope this lesson helps you gain practical knowledge in transion from virtual to physical machining. We are ready to move on to the next lesson, in which we present a case of turning a scaled baseball bat by transitioning from virtual to physical machining.

[Notes]

Lesson 13: Turning a Scaled Baseball Bat

13.1 Overview of the Lesson

This is the second application lesson. In this lesson we discuss the transition from virtual to physical machining for turning operations. We discuss a case study of turning a cylindrical steel stock for a scaled baseball bat of 12in. long to illustrate a few important factors for your consideration when you transition from virtual to physical machining. We discuss turning operations generated by SOLIDWORKS CAM and the modification of the operations before uploading the G-code post processed to a HAAS CNC lathe for turning the steel bat. We discuss the physical setup on the HAAS CNC lathe to carry out the turning operations.

Many lessons learned from Lesson 12 for milling are applicable to turning operations. Lesson 13 will be brief and will focus on turning without repeating the common factors discussed in Lesson 12.

After completing this lesson, you should acquire basic knowledge in transitioning from machining simulation to physical turning operations. In addition to pointing out important factors for your consideration, we also state a few more lessons learned that would help you avoid common pitfalls.

13.2 The Scaled Baseball Bat Solid Model

The solid model of the scaled baseball bat consists of a coordinate system (*Coordinate System1*), two solid features (*Revolve1* and *Fillet1*), and a sketch (*Sketch2*). These features are listed in the feature tree shown in Figure 13.1(a) and pointed out in the screen capture of the solid model in Figure 13.1(b).

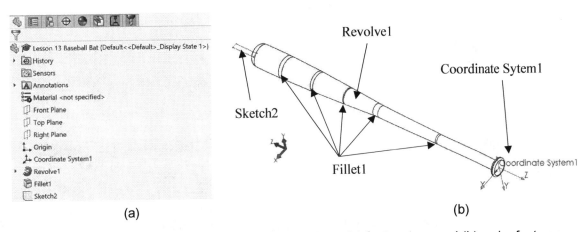

(a) (b)

Figure 13.1 The solid model of the scaled baseball bat, (a) feature tree, and (b) major features pointed out in the screen capture of the solid model

The coordinate system is defined at the right end face of the bat with its origin coinciding with the center point of the face. The Z-axis points along the longitudinal direction of the bat, which is consistent with the direction of the spindle on the CNC lathe. The X-direction coincides with the tool retract direction on the lathe.

The sketch of the revolve feature defines key dimensions of the bat. A 5in.-fillet is assigned to ensure smooth transition at the junctions where diametrical dimensions are defined in the sketch of the revolve solid feature. *Sketch2* is a simple rectangle employed for creating turning operations that cuts a groove near the left end of the stock. The groove is cut to ease the manual cut off of the bat from the stock-in-progress at the conclusion of the turning assignment.

13.3 Turning the Scaled Baseball Bat

A HAAS SL-20 lathe shown in Figure 13.2(a) is employed to machine the part. The SL-20 lathe is a single turret lathe with a work envelope ϕ15in.×21in., which is large enough to accommodate a round stock of ϕ1in.×15in. employed to turn the scaled baseball bat.

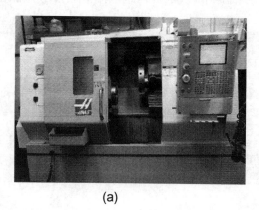 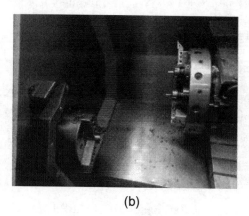

               (a)   (b)

Figure 13.2 The HAAS CNC lathe, (a) exterior view, and (b) inside view of the lathe

The low carbon alloy steel (1005) round stock of ϕ1in.×15in is clamped into a three-jar chuck of the lathe and supported by the tailstock at its right end. A machining strategy is outlined in the following steps.

Step 1. Place the left end of the round stock into a three-jar chuck of the lathe, and tailstock its right end. No face turning is required at the right end.
Step 2. Turn the round stock to the design shape of the baseball bat.
Step 3. Turn a small groove close to the left end of the stock.
Step 4. Take the stock-in-progress out of the lathe, and manually cut off the baseball bat.

13.4 Machining Operations

A round stock of ϕ1in.×15in. was created with its right end face coinciding with that of the part, as shown in Figure 13.3. As a result, no face turning at the right end face is required. You may want to open the file, *Baseball Bat with Toolpath.SLDPRT*, to review the machining operations.

Three machinable features are extracted by SOLIDWORKS CAM, *Face Feature1*, *OD feature1*, and *CutOff Feature1*, listed under SOLIDWORKS CAM feature tree shown in Figure 13.4(a). Note that *Face Feature1* is suppressed since it is not needed as discussed before. Also, *CutOff Feature1* is suppressed since a groove feature outside the left end of the bat will be turned so that the turned bat can be manually cut off easily from the stock.

Note that *OD Feature1* contains the geometric profile of the bat, as pointed out in Figure 13.4(b). *OD Feature2* that includes *Sketch2* (a small rectangle) was created to support the turn operations for cutting the groove near the left end of the stock.

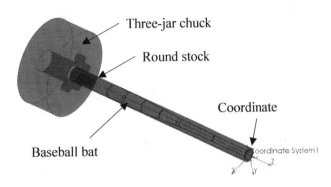

Figure 13.3 The part setup of the turning operations

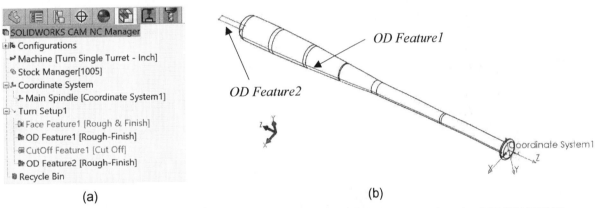

(a)   (b)

Figure 13.4 The machinable features of the baseball bat, (a) features listed under SOLIDWORKS CAM feature tree tab, and (b) turn features *OD Features1* and *2* pointed out in the solid model

Seven turning operations are created for machining the baseball bat, four rough turns and three finish turns, as listed under SOLIDWORKS CAM operation tree shown in Figure 13.5(a). To minimize tool change, all operations employ a cutter of 0.016×35° diamond insert.

The first two operations, *Turn Rough1* and *Turn Finish1*, cut the profile of the bat near its right end with a tool holder of a down left orientation, as shown in Figure 13.5(b). The material removal simulation of the two operations is shown in Figure 13.6(a).

Note that cutting the OD feature only near the right end is made by assigning adequate Z limits. This can be done through the *Advanced* tab of the *Operation Parameters* dialog box shown in Figure 13.7(a). In this case, *Z start* is set to be 0; i.e., from the right end face of the stock. The *Z end* is defined by clicking the *Select point* button ⊠—circled in Figure 13.7(a)—to bring up the *Define Point* dialog box; see Figure 13.7(b). In the *Define Point* dialog box, you may enter an offset value for Z or, in this case, pick the edge from the graphics area shown in Figure 13.7(b) as the *Z end*.

The same approach has been followed for the remaining operations, except for cutting the groove near the left end.

The next two operations, *Turn Rough2* and *Turn Finish2*, cut the main body of the bat with a tool holder of a down right orientation. The toolpaths and material removal simulation are shown in Figure 13.5(c) and Figure 13.6(b), respectively.

Turn Rough3 cuts the groove near the left end of the stock with a tool holder of a down right orientation. The toolpath and material removal simulation are shown in Figure 13.5(d) and Figure 13.6(c), respectively.

The last two operations, *Turn Rough4* and *Turn Finish4*, cut the round shape profile at the left end of the bat with a tool holder of a down right orientation. The toolpaths and material removal simulation are shown in Figure 13.5(e) and Figure 13.6(d), respectively.

13.5 Modifications of SOLIDWORKS CAM Operations

The NC operations discussed above were not automatically created by SOLIDWORKS CAM. As expected, the machining operations automatically generated by SOLIDWORKS CAM do not always work without changes. In this section, we point out changes made in the turning operations initially generated by SOLIDWORKS CAM. These turning operations generated by SOLIDWORKS CAM before modifications can be found in the file *Baseball Bat with SOLIDWORKS CAM Tooplath.SLDPRT*.

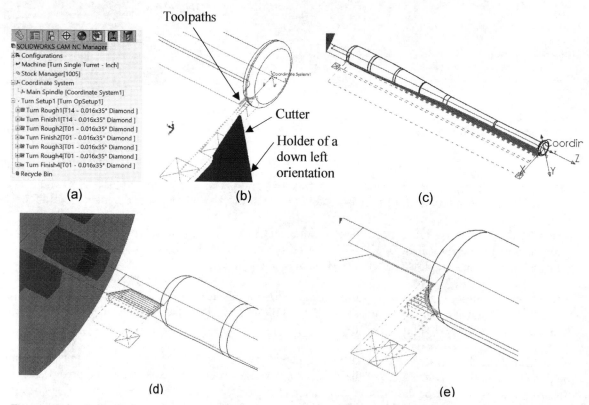

Figure 13.5 The turning operations, (a) operations listed under SOLIDWORKS CAM operation tree tab, (b) toolpaths of *Turn Rough1* and *Turn Finish1*, (c) toolpaths of *Turn Rough2* and *Turn Finish2*, (d) toolpath of *Turn Rough3*, and (e) toolpaths of *Turn Rough4* and *Turn Finish4*

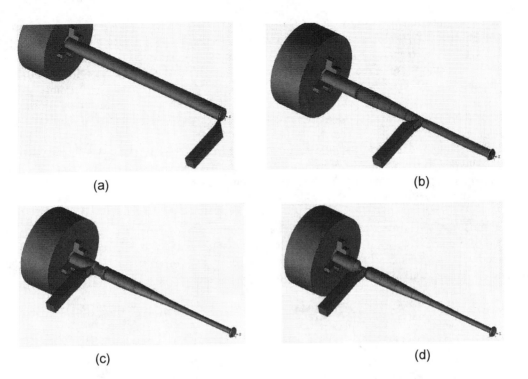

(a)

(b)

(c)

(d)

Figure 13.6 The material removal simulation of turning operations, (a) *Turn Rough1* and *Turn Finish1*, (b) *Turn Rough2* and *Turn Finish2*, (c) *Turn Rough3*, and (d) *Turn Rough4* and *Turn Finish4*

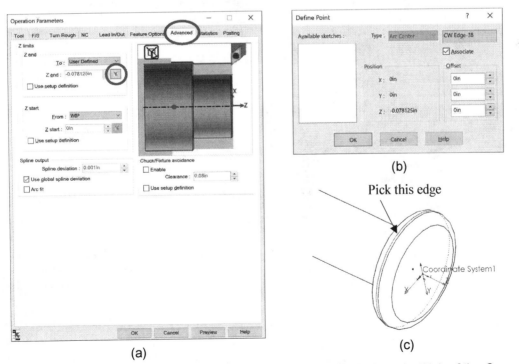

(a)

(b)

(c)

Figure 13.7 Defining the Z limits for *Turn Rough1*, (a) the *Advanced* tab of the *Operation Parameters* dialog box, (b) the *Define Point* dialog box, and (c) picking the edge from the solid model for *Z end*

There are three machinable features, *Face Feature1*, *OD Feature1*, and *CutOff Feature1*, automatically extracted by SOLIDWORKS CAM, as listed under SOLIDWORKS CAM feature tree; see Figure 13.8(a). Five turning operations are automatically generated, as shown in Figure 13.8(b). The first two face turning operations are in magenta color, indicating they are not completely defined. This is because the right end face of the stock coincides with that of the part; hence, no face turning is needed as discussed before. The next two operations, *Turn Rough1* and *Turn Finish1*, cut the entire profile of the bat using a tool holder of a down left orientation. As a result, an area close to the right end of the bat remains uncut, as pointed out in Figure 13.8(c). Such an area is not reachable by a cutter held by a tool holder of a down left orientation. *Cut Off1* operation cuts the bat off from the stock. The toolpath is shown in Figure 13.8(d). Note that the retract and approach points of the three operations determined by SOLIDWORKS CAM may be excessive. There may not be enough room to retract the cutters inside the CNC lathe when cutting the bat physically.

Although the operations seem to be working fine in the virtual environment, except for the small uncut area pointed out in Figure 13.8(c), they are not producing a desirable part. Changes are made to align the operations with the machining strategy discussed above, in addition to fixing the problems of the uncut area and excessive room required for cutter to approach and retract.

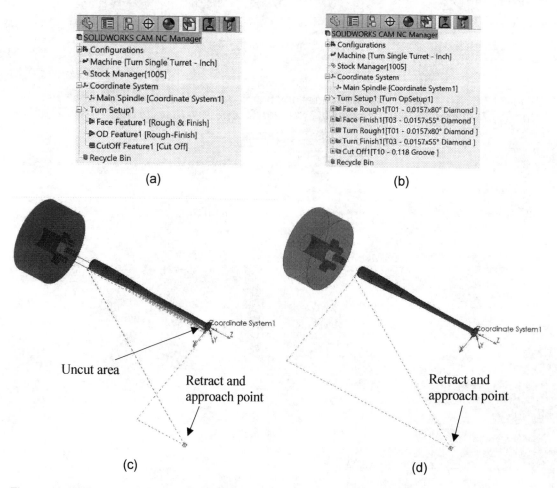

(a)   (b)

(c)   (d)

Figure 13.8 The turning operations before modifications, (a) machinable features listed under SOLIDWORKS CAM feature tree tab, (b) operations listed under SOLIDWORKS CAM operation tree tab, (c) toolpaths of *Turn Rough1* and *Turn Finish1*, and (d) toolpaths of *CutOff1*

First, a sketch of the small rectangle was added outside the left end of the bat. This rectangle defines the size of the groove that needs to be cut before manually cutting the bat off from the stock. Recall that *Turn Rough3* operation cuts the groove, as shown in Figure 13.5(d).

Second, *Face Feature1* and *Cut Off Feature1* are suppressed since they are not needed as discussed before.

Third, Z limits are added to split the profile of the bat into three segments. The first segment that covers the right end of the bat is cut by moving the cutter approaching from the right as defined in *Turn Rough1* and *Turn Finish1* operations shown in Figure 13.5(b). The second segment represents the major portion of the bat profile, which is cut by operations *Turn Rough2* and *Turn Finish2* shown in Figure 13.5(c). *Turn Rough3* shown in Figure 13.5(d) cuts a small groove close to the left end of the stock so that the turned bat can be manually cut off easily from the stock. The last segment represents the round shape of the profile at the left end of the bat, which is cut by operations *Turn Rough4* and *Turn Finish4* shown in Figure 13.5(e). Note that the main body and last segment of the bat are cut by moving the cutting approaching from the left as desired. As a result, the uncut area can now be reached. This is how the problem of the uncut area is fixed.

Last, the retract and approach points are modified to reduce cutter travel and avoid collisions. This is accomplished by using the *Operation Parameters* dialog box (see Figure 13.9). We choose *NC* tab, and enter offsets for *X reference* and *Z reference* under both *Approach* and *Retract* so that the *X* and *Z* values are close to being 1 and 0, respectively.

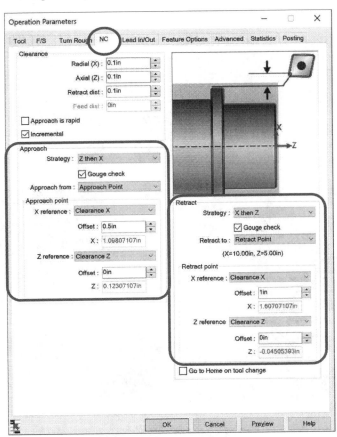

Figure 13.9 Modifying the approach and retract points

13.6 Post-Processing and G-code Modification

After making the adjustment and changes, we convert the toolpaths to G-code using a post process, TURN/HAAS_ST20, that came with SOLIDWORKS CAM, as shown in Figure 13.10.

Since a HAAS post processor is provided, the G-code converted was mostly compatible with the HAAS machines. Only minimum change for syntax was necessary, such as adding % sign at the beginning and end of the NC program, and removing N3 B90 block, as highlighted in Figure 13.11.

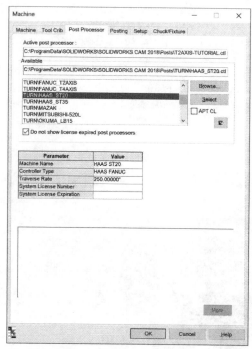

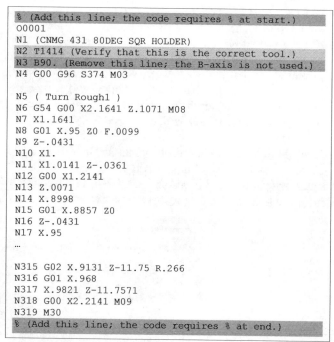

```
% (Add this line; the code requires % at start.)
O0001
N1 (CNMG 431 80DEG SQR HOLDER)
N2 T1414 (Verify that this is the correct tool.)
N3 B90. (Remove this line; the B-axis is not used.)
N4 G00 G96 S374 M03

N5 ( Turn Rough1 )
N6 G54 G00 X2.1641 Z.1071 M08
N7 X1.1641
N8 G01 X.95 Z0 F.0099
N9 Z-.0431
N10 X1.
N11 X1.0141 Z-.0361
N12 G00 X1.2141
N13 Z.0071
N14 X.8998
N15 G01 X.8857 Z0
N16 Z-.0431
N17 X.95
...

N315 G02 X.9131 Z-11.75 R.266
N316 G01 X.968
N317 X.9821 Z-11.7571
N318 G00 X2.2141 M09
N319 M30
% (Add this line; the code requires % at end.)
```

Figure 13.10 Selecting HAAS_ST20 for post processor

Figure 13.11 NC program with changes highlighted

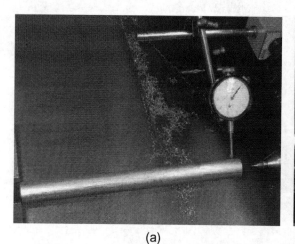

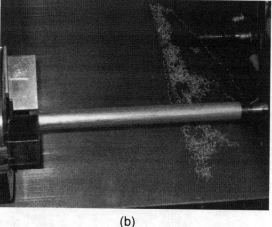

(a)   (b)

Figure 13.12 Clamping the round stock to the lathe, (a), level test, and (b) stock fully constrained with chuck and tailstock

13.7 Turning on a HAAS CNC Lathe

When the round stock was clamped to the chuck on the lathe, a pressure sensor was used to ensure that the stock was evenly inserted; i.e., the axis of the stock aligned with that of the spindle. The stock was slightly adjusted until a small variation in pressure was seen in the pressure reading, as shown in Figure 13.12(a). This is commonly referred to as level test. Once the stock was adjusted to a desired level, it was firmly clamped to the chuck and tailstocked to its right end, as shown in Figure 13.12(b).

Many of the steps involved in using the CNC lathe closely resemble the steps for the CNC mill. For example, zeroing the part is similar to zeroing parts on the mill, except that only two axes are needed. Program zero is set at the center of the right end face of the stock, corresponding to the part setup origin chosen in SOLIDWORKS CAM.

Cutting parameters were chosen based on the cutters available, the material being machined (stainless steel), and recommendations of experienced machinists at the shop floor. Surface speed, corresponding to the number of feet per minute traveled by a location at the surface of the part (contrasted with spindle speed, measured in RPM) was set at 200 ft/min. The feedrate was chosen to be 0.006 in/rev, and the maximum step depth for turn roughing was set at 0.04 in. Both right-hand and left-hand cutters were used as discussed before.

The machining time was roughly 20 minutes. Images of the physical turning process are shown in Figure 13.13(a). The machined surface, while adequate, could be improved by using dedicated finishing cutters, which have a smaller radius at the tip. The turned stock-in-progress shown in Figure 13.13(b) was removed from the lathe. The bat was separated from the stock-in-progress by manually cutting off at the right edge of the groove on the left end of the stock-in-progress.

13.8 Lessons Learned

By transitioning from virtual to physical machining discussed in this lesson, we note several important factors to take into consideration, including:

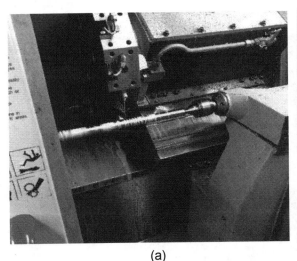

(a)   (b)

Figure 13.13 The turning operations carried out on a HAAS CNC lathe, (a) machining operations in progress, and (b) the final stock-in-progress

(1)   Similar to that of Lesson 12, it is indispensible for the CAM software users to develop a turning strategy and create virtual turning operations that support the implementation of the strategy at the shop floor.

(2)   Mounting a round stock on a CNC lathe is less involved. However, the stock may have to be adjusted to ensure that it is properly aligned with the spindle axis.

(3)   Similar to that of Lesson 12, turning operations automatically generated by SOLIDWORKS CAM may not be suitable and may not support the machining strategy developed. Turning operations and associated toolpaths automatically generated by SOLIDWORKS CAM (and other CAM software tools) must be examined carefully to verify if they are viable physically. Usually, changes are made to support an accurate and efficient machining assignment.

(4)   Again, in many situations, G-code converted from a chosen post processor of SOLIDWORKS CAM must be verified carefully to ensure the syntax and semantics are compatible with the CNC lathe.

We have completed this lesson. We hope this lesson helps you gain practical aspects in transitioning from virtual to physical turning. We are ready to move on to the final lesson of the book, in which we present additional third-party CAM modules fully integrated into SOLIDWORKS. These third-party CAM modules support complex CNC machining simulation and toolpath generations, such as machining contour surfaces of a die or mold.

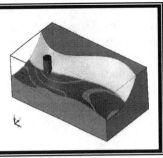

Lesson 14: Third-Party CAM Modules

14.1 Overview of the Lesson

In this lesson, we introduce third-party CAM modules that are fully integrated with SOLIDWORKS, including CAMWorks, HSMWorks, and Mastercam for SOLIDWORKS. Similar to SOLIDWORKS CAM, you may access CAMWorks, HSMWorks, and Mastercam for SOLIDWORKS from within SOLIDWORKS and create machining operations, including milling, turning, EDM, and mill-turn, to machine geometric features created in SOLIDWORKS solid models. All three CAM modules are powerful, capable of supporting almost all machining operations, and are well accepted by users in industry and academia.

SOLIDWORKS CAM is powerful, useful, and is capable of supporting many machining applications, including milling and turning. However, SOLIDWORKS CAM is somewhat limited. As one example we learned in Lesson 6, SOLIDWORKS CAM does not support contour surface milling operations. As a result, freeform surfaces often found in die and mold manufacturing are difficult to handle for a satisfactory surface finish. As we learned in Lesson 6, parts with freeform surfaces, such as the surfaces in the cavities of molds and dies, cannot be cut to meet desired surface finish requirements using SOLIDWORKS CAM. In general, SOLIDWORKS CAM does a good job in supporting machining operations for 2.5-axis features. More advanced capabilities, such as multi-axis milling, 4-axis turning, and mill-turn operations, are not supported in the current version of SOLIDWORKS CAM, even the Professional version.

More advanced and well-established CAM software tools, such as CAMWorks, Inventor HSM, and Mastercam, offer seamless integration with CAD software like SOLIDWORKS. These software tools offer a more complete suite of capabilities that support a broad range of machining operations. They take full advantage of the rich solid modeling capabilities offered in CAD. It would be beneficial to the readers in learning such third-party CAM modules that are fully integrated with SOLIDWORKS and becoming familiar with important machining capabilities beyond those of SOLIDWORKS CAM.

Among the three CAM modules included in this lesson, CAMWorks, which is an HCL Technologies Product, offers a parametric, solid-based CNC programming software system, very similar (almost identical) to and yet much more powerful than SOLIDWORKS CAM. In fact, SOLIDWORKS CAM is powered by CAMWorks. CAMWorks supports 2.5-axis to 5-axis milling, 2- and 4-axis turning, mill-turn, and multi-axis machining operations.

HSMWorks is an Autodesk Inventor CAM software completely integrated into SOLIDWORKS that streamlines the machining workflow with CAD-embedded 2.5-axis to 5-axis milling, turning, and mill-turn capabilities. HSMWorks may be interesting to readers in academia who are eligible to acquire a free software license for educational purposes. Readers may download HSMWorks from www.autodesk.com/products/hsm/overview, install the software to a computer where SOLIDWORKS resides, and go over the steps for machining a freeform surface in Lesson 6 with better machining operations to be presented in Section 14.4.

Mastercam, offered by CNC Software, Inc (http://www.mastercam.com), is considered one of the most widely used CAM software tools. Mastercam for SOLIDWORKS supports users to conduct virtual machining directly in SOLIDWORKS, using toolpaths and machining strategies preferred and prescribed by shops. With Mastercam toolpaths integrated directly within the SOLIDWORKS environment, the toolpaths are applied directly to the solid part or assembly. Any design change that Mastercam for SOLIDWORKS encounters is handled quickly, with the effected toolpaths identified so that the user can simply rebuild them. Mastercam for SOLIDWORKS supports 2.5-axis to 5-axis milling and turning operations.

We hope you have access to at least one of the three third-party CAM modules mentioned above—at least an educational copy of HSMWorks. If not, the same concept and principles discussed in this lesson may apply to other third-party CAM modules that you may have access to.

After going over this lesson, you will learn to use third-party CAM modules integrated with SOLIDWORKS other than SOLIDWORKS CAM. You will learn to use 3- or 5-axis mill for applications such as molds or die manufacturing, which involves machining a freeform surface with the final machined surface meeting the surface finish requirements.

14.2 The Freeform Surface Milling Example

In Lesson 6, we learned to use a 3-axis mill to cut a freeform surface, shown in Figure 14.1(a), by selecting a default machining strategy: *Area Clearance, Z Level*. As discussed in Lesson 6, *Area Clearance, Z Level* led to two machining operations, volume milling (*Area Clearance*) and local milling (*Z Level*).

As discussed in Lesson 6, the *Z Level* operation performs a local milling that removes material by making a series of horizontal planar cuts. The cuts follow the contour of the freeform surface at decreasing Z levels. That is, at each level, the Z-coordinate is fixed and the cutter moves in X- and Y-axes simultaneously. Therefore, *Z Level* operation is not a true contour surface milling operation, in which cutter moves in all three axes simultaneously. The surface of the stock-in-progress after *Z Level* operation is still staircase-like.

Without a contour surface milling operation, the surface finish of the freeform surface is less desired. Furthermore, due to a large curvature variation of the freeform surface in the solid model, a ball-nose cutter on a 3-axis mill is not able to reach certain areas for an accurate cut; for example, the area circled in Figure 14.1(c). A contour surface milling operation of 5-axis is required to properly machine the freeform surface.

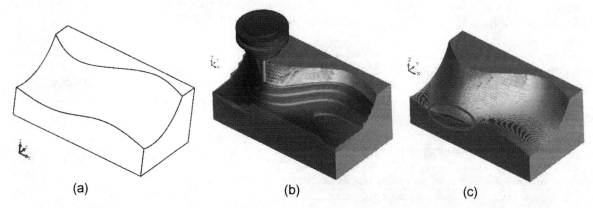

(a)   (b)   (c)

Figure 14.1 The freeform surface example, (a) part solid model, (b) material removal simulation in progress using SOLIDWORKS CAM, and (c) material remaining on the machined surface

In this lesson, we revisit this example by creating a multi-axis surface milling operation offered in other CAM modules integrated with SOLIDWORKS, including CAMWorks, HSMWorks, and Mastercam for SOLIDWORKS. A multi-axis surface milling operation cleans up the material remaining on the freeform surface and creates a much more desirable surface finish.

14.3 Removing SOLIDWORKS CAM

We first remove SOLIDWORKS CAM from the add-ins using the *Add-Ins* dialog box (see Figure 14.2). You may start SOLIDWORKS and choose from the pull-down menu

Tools > Add-Ins

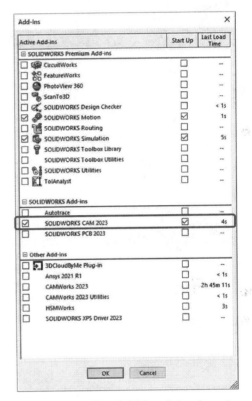

In the *Add-Ins* dialog box shown in Figure 14.2, click *SOLIDWORKS CAM* in both checkboxes (*Active Add-ins* and *Start Up*) to deselect them, and then click *OK*. The SOLIDWORKS CAM tab above the graphics area disappears momentarily.

Close SOLIDWORKS. We will start SOLIDWORKS to add in a third-party CAM module.

14.4 CAMWorks

CAMWorks was the first SOLIDWORKS® Certified Gold Product for CAM software seamlessly integrated into SOLIDWORKS. The look and feel of CAMWorks is almost identical to SOLIDWORKS CAM. In fact, SOLIDWORKS CAM is powered by CAMWorks. You may envision SOLIDWORKS CAM (even Professional version) as a stripped version of CAMWorks. Therefore, learning CAMWorks is more like continuing learning SOLIDWORKS CAM.

From SOLIDWORKS, you may open a part file with machining operations created and saved in SOLIDWORKS CAM, and enter CAMWorks to add new machining operations, or revise and delete existing operations.

Figure 14.2 The *Add-Ins* dialog box: to remove SOLIDWORKS CAM

In this section, we open the freeform surface example saved in Lesson 6, filename: *SOLIDWORKS CAM Freeform Surface with Toolpath.SLDPRT*. We will add a multi-axis surface milling operation using CAMWorks. In addition, we use the Machine Simulation capability in CAMWorks to conduct material removal simulation in a more realistic setup. Before using CAMWorks, we will have to add the module into SOLIDWORKS using the *Add-Ins* dialog box.

Add-in CAMWorks

Start SOLIDWORKS and choose from the pull-down menu

Tools > Add-Ins

In the *Add-Ins* dialog box shown in Figure 14.4, click *CAMWorks 2023* in both boxes (*Active Add-ins* and *Start Up*), and then click *OK*.

You may open the solid model, filename: *CAMWorks Freeform Surface with Toolpath.SLDPRT* to preview toolpaths to be created in this lesson. You should see the *CAMWorks 2023* tab appear above the graphics area like that of Figure 14.5 and CAMWorks tree tabs appear at the top of the feature manager window.

If you do not see any of the CAMWorks tree tabs on top of the feature manager window or any of the CAMWorks command buttons above the graphics area, you may not have set up your CAMWorks license option properly. To check the CAMWorks license setup, click the help button ⓘ (circled in Figure 14.5) and select

CAMWorks 2023 > License Info

In the *CAMWorks License Info* dialog box (Figure 14.3), click all clickable boxes (or select modules as needed, for example, *3X Mill L1* for 3-axis mill level 1) to activate the module(s). Then click *OK*.

You may need to restart SOLIDWORKS to activate CAMWorks modules newly selected from the *CAMWorks License Info* dialog box. Certainly, before going over this tutorial lesson, you are encouraged to consult with your system administrator to make sure SOLIDWORKS and CAMWorks have been properly installed on your computer.

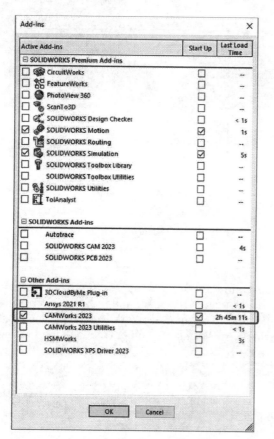

Figure 14.4 The *Add-Ins* dialog box

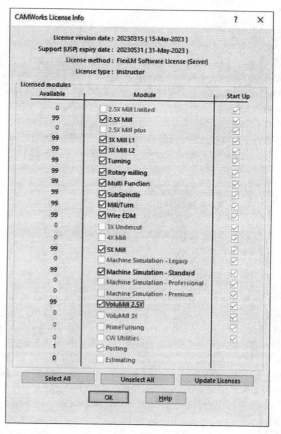

Figure 14.3 The *CAMWorks License Info* dialog box

User Interface

The overall design of CAMWorks user interface, as shown in Figure 14.5, is almost identical to that of SOLIDWORKS CAM. One minor difference that is worth pointing out is the feature tree tab icon. The third tab from the right, SOLIDWORKS CAM feature tree tab of SOLIDWORKS CAM is now replaced by the CAMWorks feature tree tab icon in CAMWorks. If you have opened the solid model, filename: *CAMWorks Freeform Surface with Toolpath.SLDPRT*, you may click CAMWorks operation tree tab , select an operation to see the corresponding toolpath in the graphics area, or click the *Simulate Toolpath* button above the graphics area to carry out material removal simulation for one or multiple operations.

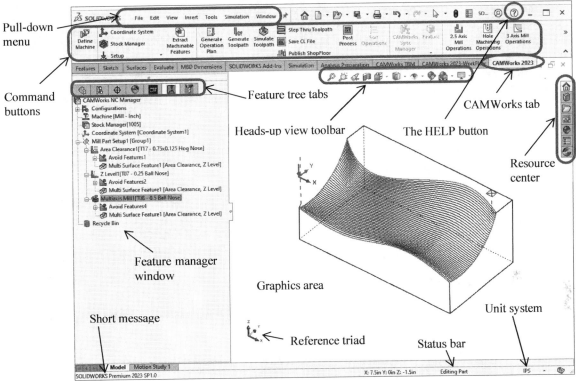

Figure 14.5 User interface of CAMWorks

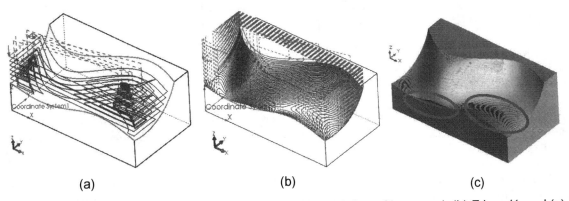

(a)   (b)   (c)

Figure 14.6 Toolpath and material removal simulation, (a) *Area Clearance1*, (b) *Z Level1*, and (c) material removal simulation of the combined two operations

We are now moving on with creating a multi-axis surface milling operation for machining the freeform surface using CAMWorks.

The Freeform Surface Example of SOLIDWORKS CAM

Recall that there were two operations we created in Lesson 6. The first operation is *Area Clearance1*, which is a rough cut (or volume milling) using a 0.75in. hog nose mill with 0.125in. corner radius (T17 in tool crib). The toolpath of the *Area Clearance1* operation is shown in Figure 14.6(a). The second operation is *Z Level1* that continues removing the material remaining from *Area Clearance1* using a smaller ball-nose cutter of diameter 0.25in. (T07); see toolpath in Figure 14.6(b).

Note that as we observed in Lesson 6, the cut was not clean, in particular, the area close to the lower edge of the freeform surface, as well as the area to the right. Both are circled in Figure 14.6(c). Moreover, staircase-like scallops remaining on the freeform surface is not desirable.

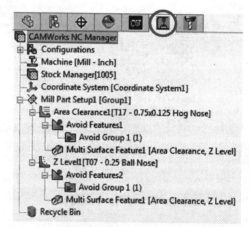

Figure 14.7   The NC operations listed under the CAMWorks operation tree tab

Open SOLIDWORKS Part

Open the part file (filename: *SOLIDWORKS CAM Freeform Surface with Toolpath.SLDPRT*) if you have not. The file contains machining operations created from Lesson 6 using SOLIDWORKS CAM. When you open the file, you should see the two operations listed under the CAMWorks operation tree tab ![icon], as shown in Figure 14.7.

5-Axis Surface Milling Operation

We now add a multi-axis surface milling operation to polish the freeform surface using a 0.5in. ball-nose cutter.

Right click *Mill Part Setup1* and choose *Multiaxis Mill Operations > Multiaxis Mill*.

The *New Operation: Area Clearance* dialog box—Figure 14.8(a)—appears. Under the *Tool* tab, select *T08 – 0.5 Ball Nose*. Click the *Features* tab, select *Multi Surface Feature1 [Area Clearance, Pattern Project]*, and click the checkmark ✔ to accept the new operation; see Figure 14.8(b).

A new operation, *Multiaxis Mill1*, is now listed in the CAMWorks feature tree tab ![icon] (see Figure 14.9) and the *Operation Parameters* dialog box appears. In the *Operation Parameters* dialog box, a 0.5in. ball-nose cutter (ID: 115, 1/2 CRB 4FL BM 1 LOC) is selected.

Click the *Pattern* tab, as shown in Figure 14.10(a), to review the pattern type and stepover. Change the pattern to *Flowline Between Curves*, and select *Zigzag* for *Pattern* under *Direction*, circled in Figure 14.10(a).

Click *Upper* in the *Operation Parameters* dialog box under the *Pattern* tab, as circled in Figure 14.10(a), the *Curve Wizard* dialog box shown in Figure 14.10(b) appears. Pick the top edge of the freeform surface; see Figure 14.10(c) in the graphics area. Notice that the point at the left end of the edge is highlighted, indicating the start point of the curve. Click the checkmark ✔ to accept the curve.

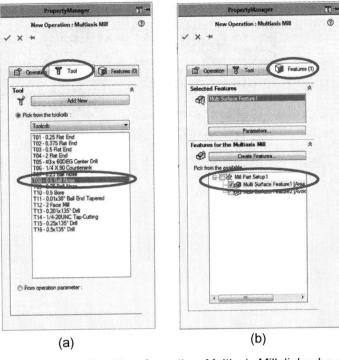

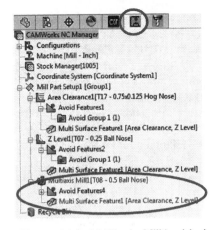

Figure 14.9 *Multiaxis Mill1* added to the feature tree

(a)   (b)

Figure 14.8 The *New Operation: Multiaxis Mill* dialog box

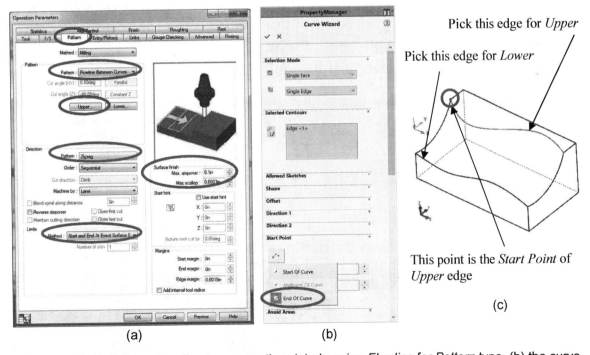

(a)   (b)

Figure 14.10 Defining a *Flowline* type operation, (a) choosing *Flowline* for *Pattern* type, (b) the curve wizard showing Edge <1> selected (in this case, the upper curve of the surface picked), and options of selecting start point, and (c) picking edges of the freeform surface from the graphics area

Click *Lower*, and pick the lower edge of the freeform surface in the graphics area. Notice that the point at the left end of the curve is highlighted, indicating the start point of the curve. The start points of the two edges are consistent, which leads to a valid toolpath. No change is necessary. Note that if this is not the case, you need to click the *End of Curve* button, circled in Figure 14.10(b), to set the left end point of the curve as the start point.

Choose S*tart and End At Exact Surface Edge* for *Method* under *Limits* in the *Operation Parameters* dialog box under the *Pattern* tab, as circled in Figure 14.10(a). Change the *Max. scallop* to *0.0051in*, as shown in Figure 14.10(a).

Click the *Axis Control* tab (see Figure 14.11). Make sure that *5 Axis* is chosen, and select *Normal to Surface* for *Tool axis*.

Click the *Preview* button to show the toolpath like that of Figure 14.12(a). We accept the toolpath.

Click the close button ⊠ at the top right corner of the *Operation Parameters* dialog box, and click *OK* to accept the changes and regenerate the toolpath.

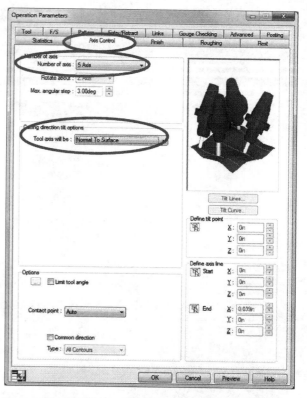

Figure 14.11 Selecting tool axis control

Simulate the toolpaths by clicking the *Simulate Toolpath* button 🔧 Simulate Toolpath above the graphics area.

Click the *Run* button ▶ to simulate the toolpath. The material removal simulation of the operations appears in the graphics area, similar to that of Figure 14.12(b). The surface finish is clean and desirable. No uncut material remains, and no staircase scallop exists. We rename the operation as *Multi-axis Surface Milling*.

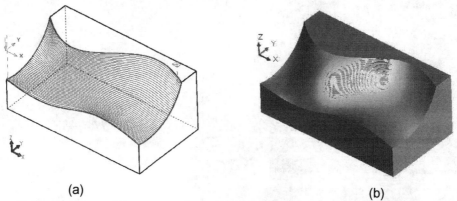

(a)   (b)

Figure 14.12 Multi-axis surface milling operation created in CAMWorks, (a) toolpath, and (b) the material removal simulation of all three operations combined, *Area Clearance1*, *Z Level1*, and *Multi-axis Surface Milling*

We have completed the multi-axis milling operation in CAMWorks. You may save your model for future reference. Next, we take a look at Machine Simulation capability of CAMWorks, which offers a more realistic material removal simulation.

Machine Simulation

Machine Simulation in CAMWorks offers a more realistic setup in simulating machining operations. For example, a sample machine simulator called *Mill_Tutorial* (among others provided by CAMWorks such as *MillTurn_Tutorial*), provides a setup of tools with a tool holder and stock mounted on a tilt-rotary table, as shown in Figure 14.13.

Machine Simulation also detects tool or tool holder collisions when they happen. Note that you may add your own virtual CNC machines, resembling the physical machines at your machine shop, to CAMWorks that allow you to carry out machining simulation in a much more realistic setting. You may also add associated post processors to CAMWorks that support generating G-code compatible with the respective CNC machines.

To bring up the Machine Simulation window, you may right click *Mill Part Setup* under CAMWorks operation tree 🔲, and choose *Machine Simulation* and then *Legacy*.

The layout of the *Machine Simulation* window is shown in Figure 14.13.

In the graphics area, the cutter, tool holder, tilt table, rotary table, and the stock together with a machine coordinate system XYZ are displayed. To the right, the three machining operations and corresponding G-code are listed in the upper and lower areas of *Move List*, respectively.

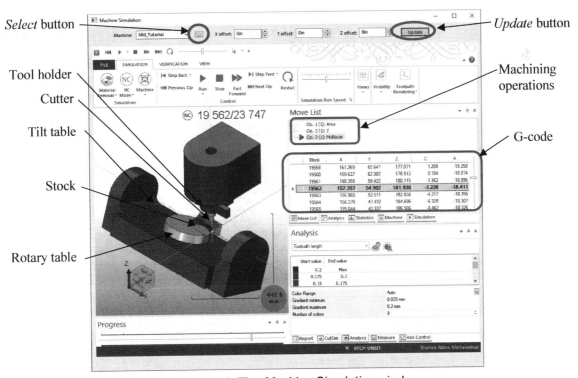

Figure 14.13 The *Machine Simulation* window

On top, the default simulator, *Mill_Tutorial*, is selected. Below are buttons that control the machining simulation run. Under these buttons are four tabs, *FILE*, *SIMULATION*, *VERIFICATION*, and *VIEW*, supporting respective uses of the machine simulation capabilities.

SIMULATION tab is chosen as default. Under the *SIMULATION* tab, there are *Simulation* options that offer setups to display the desired simulation, such as material removal. Next to the *Simulation* options are command buttons that provide options for *Control*, *Simulation Run Speed*, and *Views*.

Before we start, click the *Select* button ☐ next to the *Machine* on top of the *Machine Simulation* window (circled in Figure 14.13) to bring up the *Select Point* dialog box (Figure 14.14). We are about to select the origin of the work coordinate system on the part that determines its position and orientation with respect to the top face of the rotary table.

Pick the coordinate system of the part, *Coordinate System1*, in the graphics window (see Figure 14.15), and enter 3.75 and 2.0 for *X* and *Y* offsets to determine the origin of the work coordinate system. And then, click the *Update* button on top. The origin of the work coordinate system (at the center of the bottom face of the stock) coincides with the center point at the top face of the rotary table (see Figure 14.13).

To verify if a correct work coordinate system has been chosen for the Machine Simulation, you may choose *TableTable* as the machine by pulling down the *Machine* selection and choosing it (see Figure 14.16). Click the *Update* button. In the graphics area, a work coordinate system appears at the center of the bottom face of the stock with its X-, Y-, and Z-axes parallel to those of the virtual machine shown at the lower left corner of Figure 14.17.

Now, we choose *Mill_Tutorial* as the machine and click the *Update* button, and run the simulation.

Click the *Run* button ▶ to start the simulation. The G-code of the corresponding tool motion is displayed in the lower portion of the *Move List* area. The simulation runs to the end for all three operations without any issue. There is no collision encountered.

Close the *Machine Simulation* window. Suppress the first two operations by right clicking *Area Clearance1* (and *Z Level1*) and choosing *Suppress* under the CAMWorks operation tree ▣.

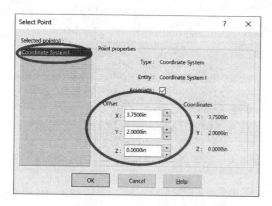

Figure 14.14 The *Select Point* dialog box

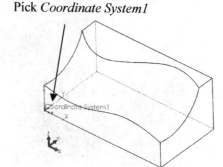

Pick *Coordinate System1*

Figure 14.15 Picking the coordinate system, *Coordinate System1*

Figure 14.16 Choosing *TableTable* for machine

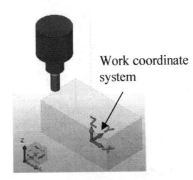

Figure 14.17 The work coordinate system displayed

Rerun the machine simulation. The simulation of the third operation, *Multiaxis Surface Milling*, starts immediately since the first two operations are suppressed.

Without *Area Clearance1* and *Z Level1* operations, a collision between the tool holder and the stock is detected immediately after the simulation starts. Click *Yes to All* in the warning box, as shown in Figure 14.18(a), to allow the simulation to continue. The collision area is highlighted in red—Figure 14.18(b)—and a symbol ✸ appears in the problematic NC blocks in the G-code where collision is first detected; see Figure 14.18(c).

Close the *Machine Simulation* window. Unsuppress the first two operations by right clicking *Area Clearance1* (and *Z Level1*) and choosing *Unsuppress* under the CAMWorks operation tree 🖳 .

We have completed the exercise. You may save the model under a different name for future reference. Close SOLIDWORKS.

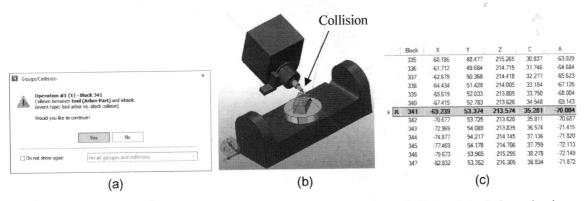

Figure 14.18 The collision detection in *Machine Simulation*, (a) the *Collision detected* warning box, (b) collision area highlighted, and (c) the symbol ✸ appearing in front of problematic NC blocks

14.5 HSMWorks

Inventor HSMWorks software is a CAM add-in designed to work seamlessly with SOLIDWORKS. HSMWorks offers machining workflow with CAD-embedded 2.5-axis to 5-axis milling, turning, and mill-turn capabilities. In addition, HSMWorks supports simulation capabilities for forward and reverse material removal, customizable post processor system with industry-standard posts included, and NC program editor with capabilities of file compare, block numbering, text editing and remove comments, similar to that of SOLIDWORKS CAM NC Editor.

You may download HSMWorks for a 30-day trial (www.autodesk.com/products/hsm/overview) or sign in to receive an Educational license (www.autodesk.com/education/free-software/hsmworks-ultimate).

Either way, you should be able to go through the machining example of the freeform surface model in this section.

Once you download and install the software on your computer, you may start SOLIDWORKS and add-in HSMWorks.

Add-in HSMWorks

Start SOLIDWORKS and choose from the pull-down menu

Tools > Add-Ins

In the *Add-Ins* dialog box shown in Figure 14.19, click *HSMWorks* in both checkboxes (*Active Add-ins* and *Start Up*), and then click *OK*.

You should see that the *CAM* tab appears above the graphics area like that of Figure 14.20 (with no solid model opened yet) and an *HSMWorks* tree tab ▼ added to the top of the feature manager window.

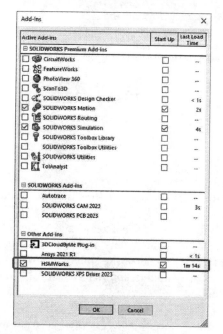

Figure 14.19 The *Add-Ins* dialog box

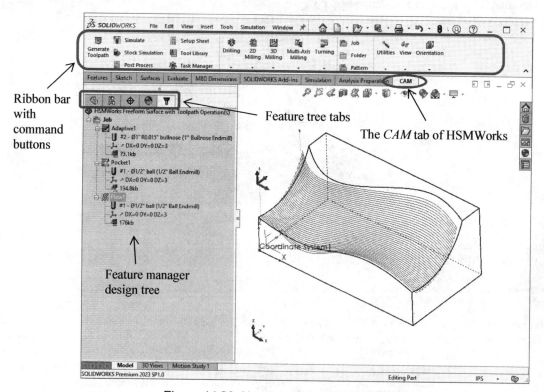

Ribbon bar with command buttons

Feature tree tabs

The *CAM* tab of HSMWorks

Feature manager design tree

Figure 14.20 User interface of HSMWorks

User Interface

Open the solid model, filename: *HSMWorks Freeform Surface with Toolpath.SLDPRT*. Browse existing operations to become familiar with the buttons and selections.

The overall design of HSMWorks user interface, as shown in Figure 14.20, is very similar to that of SOLIDWORKS CAM.

As shown in Figure 14.20, the user interface window of HSMWorks consists of pull-down menus, ribbon bar with command buttons, graphics area, and feature manager window.

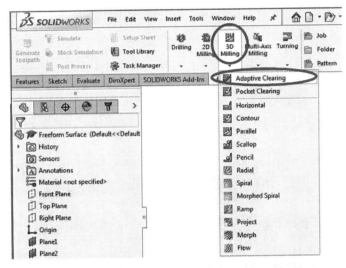

Figure 14.22 Clicking *Library* to choose a cutter

Figure 14.21 Defining an *Adaptive Clearing* operation

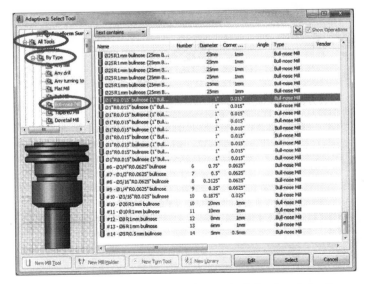

Figure 14.23 Selecting 1" bull nose mill for the *Adaptive Clearing* operation

Figure 14.24 Tool selected

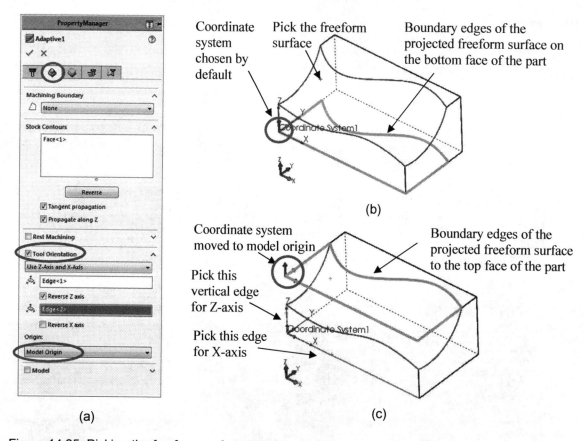

(a)

(b)

(c)

Figure 14.25 Picking the freeform surface and defining the coordinate system, (a) the *Geometry* tab, (b) picking the freeform surface from the graphics area, and (c) defining the coordinate system

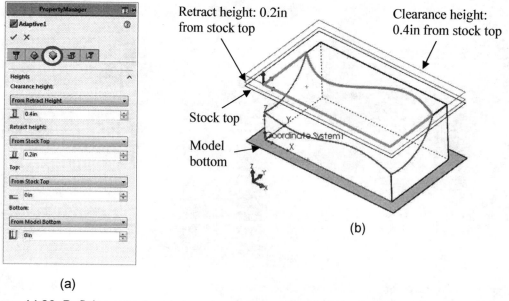

(a)

(b)

Figure 14.26 Defining retract and clearance, (a) the *Heights* tab, (b) heights shown in the graphics area reflecting the numeric figures defined in the dialog box

The three machining operations created are listed in the feature manager tree. They are *Adaptive1*, *Pocket1*, and *Flow1*. *Adaptive1* carries out volume milling using a 1in. bull nose endmill with 0.015in. corner radius similar to that of *Area Clearance* of SOLIDWORKS CAM. *Pocket1* is a local milling; and *Flow1* is a surface milling, in this case, a 3-axis surface milling operation.

Close the part file *HSMWorks Freeform Surface with Toolpath.SLDPRT* and open *Freeform Surface.SLDPRT*. We will learn to create the three operations mentioned above using HSMWorks next.

3D Milling: Adaptive

Click *3D Milling* and choose *Adaptive Clearing* (circled in Figure 14.21). In the *PropertyManager: Adaptive1* dialog box appearing next, click *Library* under the *Tool* tab 🕮 (selected by default) to select a cutter (circled in Figure 14.22).

In the *Adaptive1: Select Tool* dialog box, expand *All Tools*, expand *By Type*, choose *Bull nose Mill* and select a tool of 1" bullnose endmill with 0.015" corner radius, as shown in Figure 14.23. Click *Select* to select the tool and close the dialog box. Machining parameters, such as feedrate, spindle speed, appear (see Figure 14.24). We will use the default parameters for the operation.

Next, we click the *Geometry* tab 🔵, circled in Figure 14.25(a), and pick the freeform surface from the graphics area. An orange loop that represents the boundary edges of the projected freeform surface on the bottom face of the part appears; see Figure 14.25(b).

The fact that we see the freeform surface projected to the bottom face of the part is because the origin of the default coordinate system has been chosen at the bottom left corner of the part—circled in Figure 14.25(b).

We will move the origin to the top left corner of the bounding box of the part.

Select *Tool Orientation* by clicking the small checkbox next to it—circled in Figure 14.25(a). Keep *Use Z-axis and X-axis* selected, and pick a vertical edge of the part for Z axis. Select *Reverse Z axis* if the Z-axis arrow points down. Pick a horizontal edge of the part for X-axis.

Select *Model Origin* for defining the origin of the coordinate system—circled in Figure 14.25(a). The origin of the coordinate system moves to the top left corner, which is desirable.

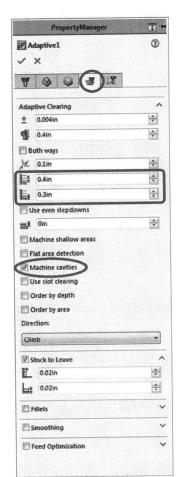

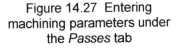

Figure 14.27 Entering machining parameters under the *Passes* tab

Figure 14.28 Machining settings under the *Linking* tab

The orange loop that represents the boundary edges of the projected freeform surface is now moved to the top face of the part; see Figure 14.25(c).

Choose the *Heights* tab 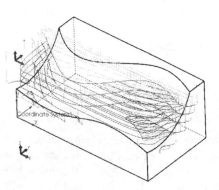—circled in Figure 14.26(a)—and use all default values for retract and clearance, as illustrated in Figure 14.26(b).

Choose the *Passes* tab (circled in Figure 14.27) to enter 0.4in. for *Maximum stepdown*, 0.3in. for *Fine stepdown*, and choose *Machine cavities*.

Choose the *Linking* tab (circled in Figure 14.28) and use all default settings.

We have now completely defined the adaptive operation. Click the checkmark ✔ to accept the definition and close the *PropertyManager: Adaptive1* dialog box.

The toolpath appears on the part in the graphics area, as shown in Figure 14.29. Note that yellow curves indicate tool retract, and blue curves are cutter locations of cutting motion.

Figure 14.29 Toolpath of the *Adaptive1* operation

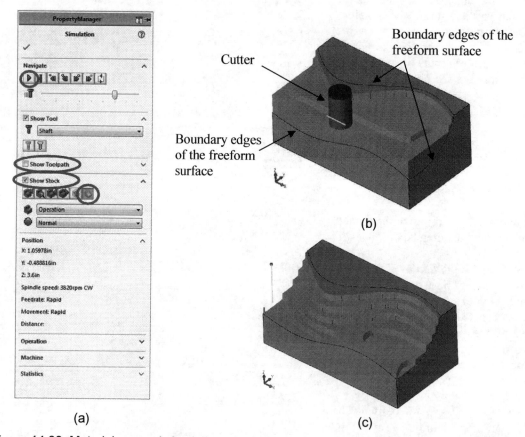

(a)

(b)

(c)

Figure 14.30 Material removal simulation of the *Adaptive1* operation, (a) the *PropertyManager: Simulation* dialog box, (b) simulation in progress, and (c) stock-in-progress at the end of the simulation

We are ready to show the material removal simulation.

Click the *Simulate* button 🌡 Simulate from the command ribbons above the graphics area.

In the *PropertyManager: Simulation* dialog box—Figure 14.30(a), we first turn off the toolpath display by clicking the checkbox in front of the *Show Toolpath*—circled in Figure 14.30(a).

We turn on stock display by clicking the checkbox in front of *Show Stock*—circled in Figure 14.30(a). Click the *Show/hide bodies* button 🔘, the right most button circled in Figure 14.30(a), to show the stock. Then, click the *Transparent Stock* button 🔘 (left of the *Show/hide bodies* button 🔘) to toggle stock display to be transparent.

Click the *Play* button ▶ to start the simulation. The stock material is removed slice by slice like a typical volume milling operation as expected; see Figure 14.30(b). A staircase surface reveals the machined surface at the end of the *Adaptive1* operation, as shown in Figure 14.30(c).

3D Milling: Pocket Rest milling

The second operation to create is a pocket rest milling, serving as a local milling to continue removing material remaining on the freeform surface using a smaller cutter, in this case, a 0.5in. ball mill.

Click *3D Milling* and choose *Pocket Clearing* (see Figure 14.31). In the *PropertyManager: Pocket1* dialog box (see Figure 14.22), click *Library* to bring out the *Pocket1: Select Tool* dialog box.

In the *Pocket1: Select Tool* dialog box (see Figure 14.32), expand *All Tools*, expand *By Type*, choose *Ball Mill* and select a tool of 0.5" ball mill. Click *Select* to select the tool and close the dialog box.

Machining parameters, such as feedrate and spindle speed, appear (similar to that of Figure 14.24). We stay with the default parameters for the operation.

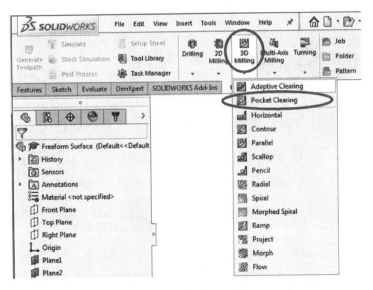

Figure 14.31 Selecting *Pocket Clearing* to create a local milling operation

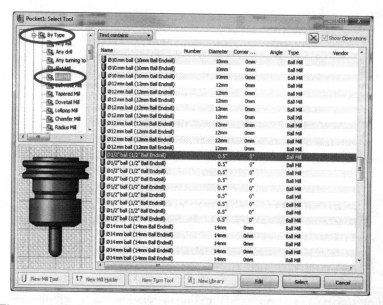

Figure 14.32 Selecting 0.5" ball mill for the *Pocket Clearing* operation

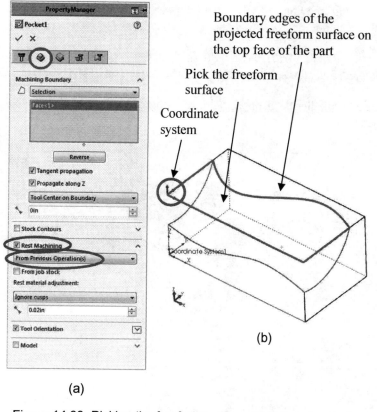

(a)

Figure 14.33 Picking the freeform surface and choosing rest machining, (a) the *Geometry* tab, and (b) pick the freeform surface from the graphics area

Boundary edges of the projected freeform surface on the top face of the part

Pick the freeform surface

Coordinate system

(b)

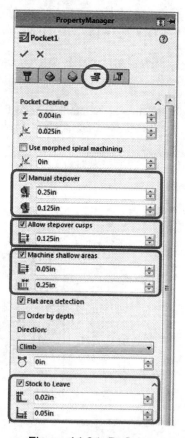

Figure 14.34 Defining machining parameters under the *Passes* tab

Next, we click the *Geometry* tab ◆, and pick the freeform surface from the graphics area (see Figure 14.33). An orange loop that represents the boundary edges of the projected freeform surface now appears on the top face of the stock, which is desirable.

Select *Rest Machining* by clicking the checkbox next to it [circled in Figure 14.33(a)], and select *From Previous Operation(s)*, also circled in Figure 14.33(a).

We use the default settings for retract and clearance.

Then, we choose the *Passes* tab ▤ (circled in Figure 14.34), choose *Manual stepover*, and enter 0.25in. for *Maximum Stepover*, 0.125in. for *Minimum Stepover*. Choose *Allow Stepover Cusp*, and enter 0.125in., as shown in Figure 14.34. Choose *Machine shallow area*, and enter 0.05in. for *Minimum Shallow Stepdown*, and 0.25in. for *Maximum Shallow Stepover*. Choose *Stock to leave*, and enter 0.02in. for wall surface and 0.05in. for bottom face.

Choose the *Linking* tab ▤ and use all default settings.

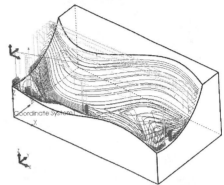

Figure 14.35 Toolpath of the *Pocket1* operation

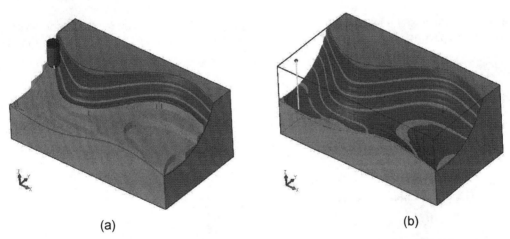

(a)   (b)

Figure 14.36 Material removal simulation of the *Pocket1* operation, (a) simulation in progress, and (b) stock-in-progress at the end of the simulation

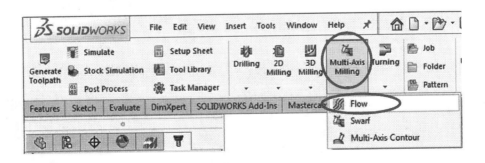

Figure 14.37 Defining a *Multi-Axis Milling: Flow* operation

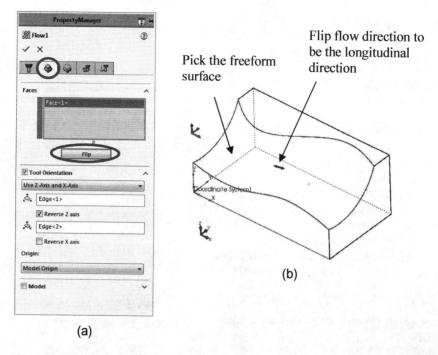

(a)

(b)

Figure 14.38 Picking the freeform surface and flipping the flow direction, (a) the *Geometry* tab circled, and (b) picking the freeform surface from the graphics area and flipping the flow direction

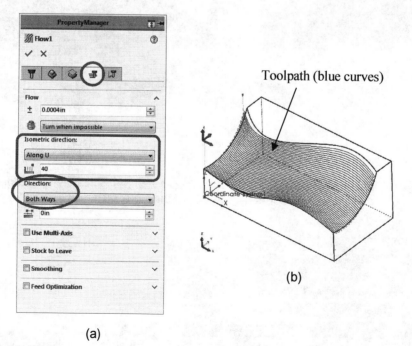

(a)

(b)

Figure 14.39 Defining parameters for the toolpath, (a) the *Passes* tab, and (b) toolpath of the multi-axis: flow operation

We have now completely defined the pocket operation. Click the checkmark ✔ to accept the definition and close the *PropertyManager: Pocket1* dialog box.

The toolpath appears in the part in the graphics area, as shown in Figure 14.35.

Click the *Simulate* button 🍅 Simulate from the command ribbons above the graphics area to show material removal simulation, as shown in Figure 14.36.

The machine surface is much cleaner. We are now ready to move on for creating a surface milling operation to polish the freeform surface.

Multi-Axis Milling: Flow

Click *Multi-Axis Milling* button ⬚ from the ribbon bar above the graphics area, and select *Flow* (circled in Figure 14.37). In the *PropertyManager: Flow1* dialog box appearing next, click *Library* to select a 0.25in. ball-nose cutter.

Next, we click the *Geometry* tab ◈, and pick the freeform surface from the graphics area. An arrow pointing along the up-down direction appears on the freeform surface. Click the *Flip* button, circled in Figure 14.38(a), to flip the flow direction to be pointing along the longitudinal direction, as shown in Figure 14.38(b).

We use the default settings for retract and clearance.

Choose the *Passes* tab ⬛, choose *Along U*, enter 40 for *Number of Stepovers*, and select *Both Ways*, as shown in Figure 14.39(a). Note that we did not select *Use Multi-Axis* since the operation is assumed 3-axis.

Choose the *Linking* tab ⬛ and use all default settings.

We have now completely defined the pocket operation. Click the checkmark ✔ to accept the definition and close the *PropertyManager: Flow1* dialog box.

The toolpath appears in the part in the graphics area, as shown in Figure 14.39(b).

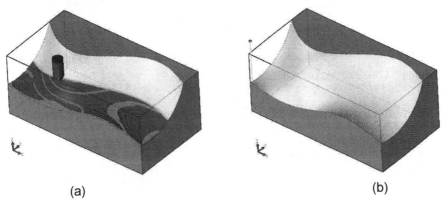

(a)   (b)

Figure 14.40 Material removal simulation of combined three operations, (a) *Flow1* simulation in progress, and (b) stock at the end of the simulation

Click the *Simulate* button 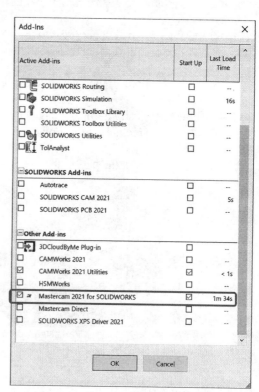 Simulate from the command ribbons above the graphics area to show material removal simulation, as shown in Figure 14.40.

The machined surface is clean; the excessive material close to the front edge of the freeform surface is removed completely. You may increase the number of passes to further reduce the scallop height of the freeform surface.

We have now completed the exercise. You may save the model under a different name for future references.

14.6 Mastercam for SOLIDWORKS

Mastercam is a computer-aided manufacturing (CAM) software program used by manufacturing professionals, such as machinists and computer numerical control (CNC) programmers.

Mastercam for SOLIDWORKS is a software module that combines the widely used CAM software with SOLIDWORKS so that users can program parts directly in SOLIDWORKS, using toolpaths and machining strategies preferred the most by shops. SOLIDWORKS users will feel at ease with the Mastercam machining tree, which delivers quick access to any point in the machining process.

Mastercam for SOLIDWORKS supports toolpath generation in part or assemblies as well as multiple configurations within each. 2D, 3D, 5 axis milling, and lathe turning are supported. In addition, Mastercam for SOLIDWORKS offers material removal verification and machine simulation (similar to that of CAMWorks) capabilities.

Once you installed the software on your computer, you may start SOLIDWORKS.

Add-in Mastercam

Start SOLIDWORKS and choose from the pull-down menu

Tools > Add-Ins

In the *Add-Ins* dialog box shown in Figure 14.41, click *Mastercam 2021 for SOLIDWORKS* in both checkboxes (*Active Add-ins* and *Start Up*), and then click *OK*. You should see that *Mastercam 2021* tab appears above the graphics area like that of Figure 14.42 (with no solid model opened yet) and a Mastercam Toolpaths Manager tab added to the top of the feature manager window.

Open the solid model file *Mastercam Freeform Surface with Toolpath.SLDPRT*. You may browse existing operations to get familiar with the buttons and selections.

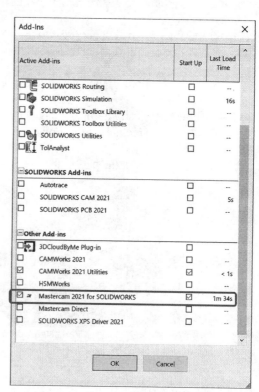

Figure 14.41 The *Add-Ins* dialog box

User Interface

The overall design of Mastercam for SOLIDWORKS user interface, as shown in Figure 14.42, is very similar to that of SOLIDWORKS CAM. As shown in Figure 14.42, the user interface window of Mastercam for SOLIDWORKS consists of pull-down menus, ribbon bar with command buttons, graphics area, and feature manager window.

The three operations created are listed in the feature manager tree. They are *1 - Surface Rough Flowline*, *2 – Surface Restmill*, and *3 – Flow 5 Axis*. *Surface Rough Flowline* carries out volume milling using a 1in. bull-nose cutter of 0.25in. corner radius like that of *Area Clearance* of SOLIDWORKS CAM. *Surface Restmill* is a local milling; and *Flow 5 Axis* is a surface milling, in this case, a 5-axis surface milling operation.

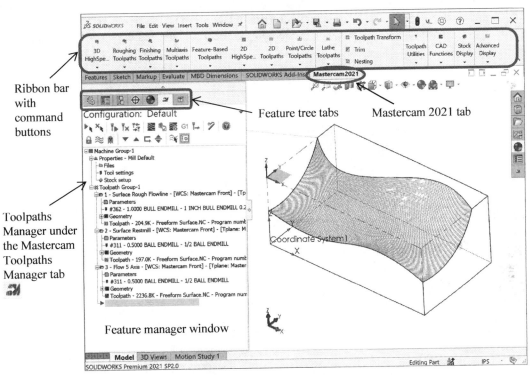

Figure 14.42 User interface of *Mastercam for SOLIDWORKS*

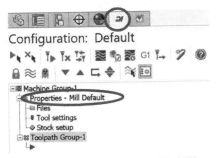

Figure 14.43 A default mill, *Mill Default*, listed in the Toolpaths Manager

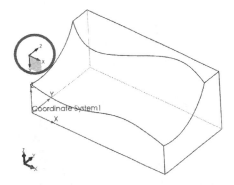

Figure 14.44 The default coordinate system assigned by Mastercam

Close the model file: *Mastercam Freeform Surface with Toolpath.SLDPRT*, and open *Freeform Surface.SLDPRT*. We will learn to create the three operations mentioned above using Mastercam next.

Reset Coordinate System

After opening the part, *Freeform Surface.SLDPRT*, we select Mastercam Toolpaths Manager tab 🚀 to display the *Toolpaths Manager*.

Mastercam for SOLIDWORKS automatically adds a default machine: *Mill Default* in the *Toolpaths Manager*, as shown in Figure 14.43.

When the part is brought into SOLIDWORKS, Mastercam assigned a default coordinate system with X axis points along the longitudinal direction; in this example, Y axis points downward, and Z axis points into the part, as shown in Figure 14.44. Apparently, the orientation of the coordinate system is not desirable and must be reset. The origin of the coordinate system is located at the top left corner of the bonding box of the part, which is desirable.

Click the *Plane Manager* button 📋, next to the Mastercam Toolpaths Manager tab 🚀 of the FeatureTree tabs (circled in Figure 14.45) at top to bring up the *Plane Manager* dialog box (see Figure 14.45).

Select *Mastercam Front* (the second row); the orientation of the coordinate system changes to that of Figure 14.46 with Z-axis pointing upwards, which is desirable.

Click the equal button = to set and accept current plane and origin.

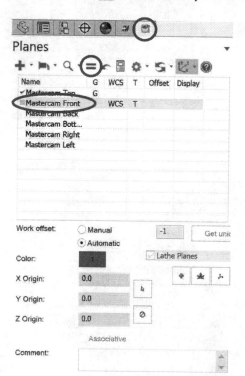

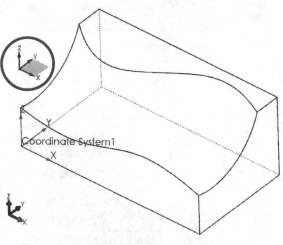

Figure 14.46 The coordinate system with desired orientation

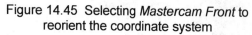

Figure 14.45 Selecting *Mastercam Front* to reorient the coordinate system

Define and Display Stock

Next, we define stock. Expand the node *Properties – Mill Default* under the *Toolpaths Manager*, and click *Stock setup*. In the *Machine Group Properties* dialog box, select *Stock Setup* tab; see Figure 14.47(a). Enter 7.5, 3.0, and 4.0 for length, width, and height respectively of the stock, and pick the top left corner of the rectangular block for *Stock Origin*. Enter 4.0 for Z under *Stock Origin*. Select *Display* to display the stock in dotted line, as shown in Figure 14.48.

Click the *Tool Settings* tab and select: *ALUMINUM inch – 2024* (default selection) for material and unit.

Click the checkmark ☑ to accept the selection and close the dialog box. We are ready to create the first milling operation.

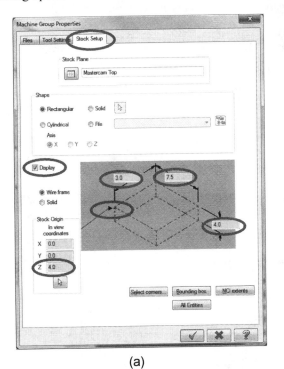
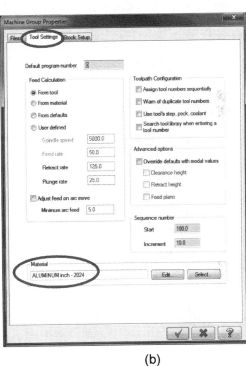

(a)   (b)

Figure 14.47 Defining stock, (a) stock size and origin under the *Stock Setup* tab, and (b) stock material and unit under the *Tool Settings* tab

Surface Rough Flowline

Click *Roughing Toolpaths* and choose *Flowline* (see Figure 14.49).

Choose *Undefined* (default) in the *Select Boss/Cavity* dialog box (see Figure 14.50), and click the checkmark ☑.

In the *PropertyManager* dialog box appearing next, we leave the *Mastercam Front* as the *Tool Plane*, and pick the freeform surface from graphics area, as shown in Figure 14.51. Click the checkmark ✓.

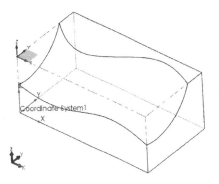

Figure 14.48 Stock shown in dotted lines

In the *Surface Rough Flowline* dialog box (Figure 14.52), click *Select library tool* to bring up the *Tool Selection* dialog box (see Figure 14.53), from which we pick Tool #362; 1" endmill bull (corner radius: 0.25"). Click the checkmark ✅ to accept the selection.

Choose the *Surface parameters* tab, and enter 0.025 for *Stock to leave on drive* (see Figure 14.54).

Choose the *Rough flowline parameters* tab (see Figure 14.55), and enter 0.5 for *Max stepdown*, 0.9 for *Stepover*, and *Zigzag* for *Cutting method* (default). Click the checkmark ✅ to accept the definition.

Flowlines along the up-down direction over the freeform surface appear in the graphics area, as shown in Figure 14.56(a). In the *Flowline options* of the *PropertyManager* dialog box—see Figure 14.56(b), click *Cut Direction* to toggle the flowlines to be along the longitudinal direction; see Figure 14.56(c).

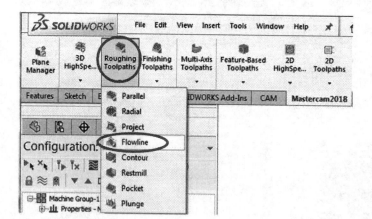

Figure 14.50 The Select Boss/Cavity dialog box

Figure 14.49 Defining *Surface Rough Flowline* operation

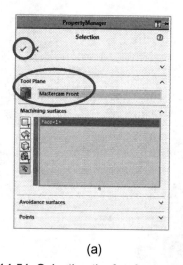

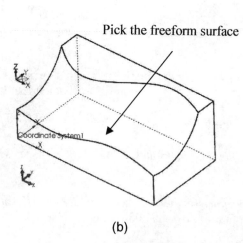

Pick the freeform surface

(a)   (b)

Figure 14.51 Selecting the freeform surface, (a) keeping the *Mastercam Front* as the *Tool Plane*, and (b) picking the freeform surface from the graphics area

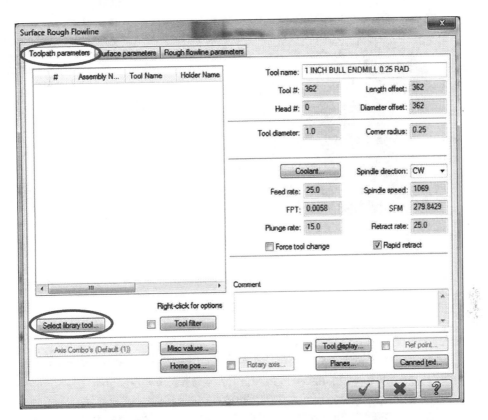

Figure 14.52 Clicking *Select library tool* from the *Surface Rough Flowline* dialog box

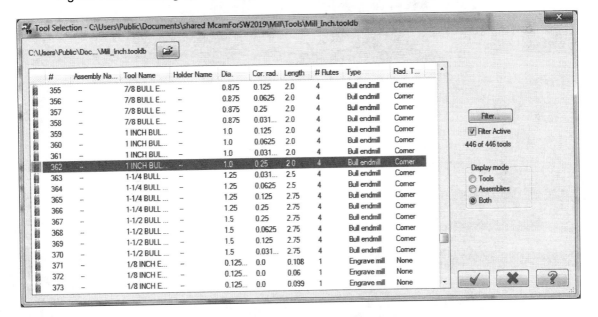

Figure 14.53 Choosing Tool #362 from the *Tool Selection* dialog box

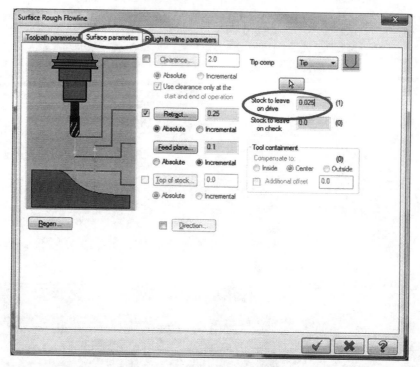

Figure 14.54 Defining stock to leave under the *Surface parameters* tab of the *Surface Rough Flowline* dialog box

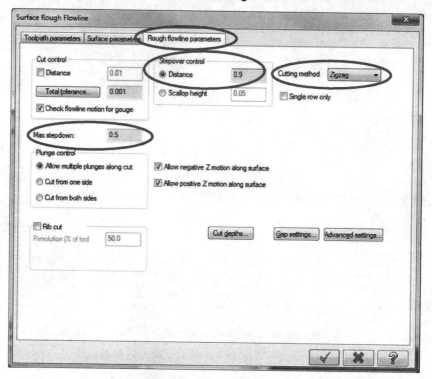

Figure 14.55 Defining cutting parameters under the *Rough Flowline parameters* tab of the *Surface Rough Flowline* dialog box

Click the checkmark ✔ at the top left corner of the *Flowline* options dialog box to accept the changes. Toolpaths will be calculated and displayed in the graphics area like that of Figure 14.56(d).

The *Surface Rough Flowline* operation is completely defined. Click *1 – Surface Rough Flowline* in the *Toolpaths Manager* (see Figure 14.57), and click the *Verified selected operations* button 🔧 above (the sixth button from the left in the first row, circled in Figure 14.57) to bring out the *Mastercam Simulator* window (Figure 14.58).

Click the *Play* button (circled in Figure 14.58) to carry out the material removal simulation. Significant amount of material remained on the freeform surface at the end of the operation. We will create a local milling to remove more material next.

Roughing Toolpaths: Restmill

Click *Roughing Toolpaths* and choose *Restmill* (see Figure 14.59).

In the *PropertyManager* dialog box appearing next, we leave the *Mastercam Front* as the *Tool Plane*, and pick the freeform surface from the graphics area, as shown in Figure 14.60. *Face<1>* is now listed in the *Machining surfaces* field.

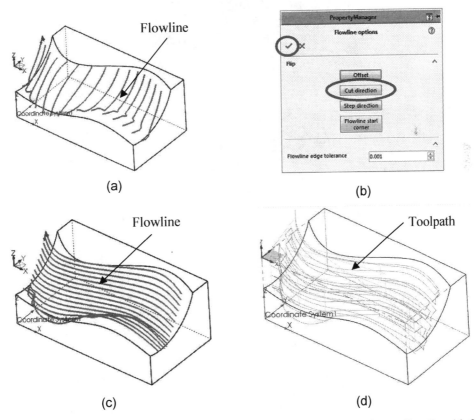

Figure 14.56 Defining flowline (or cut) direction, (a) flowlines in the up-down direction (default), (b) the *Flowline options* dialog box, (c) flowlines in the longitudinal direction, and (d) the toolpaths generated for the *Surface Rough Flowline* operation

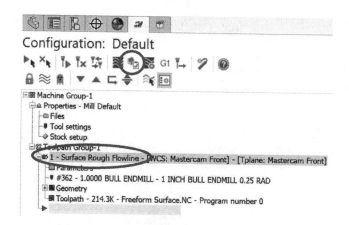

Figure 14.57 The *1 – Surface Rough Flowline* listed in the *Toolpaths Manager*

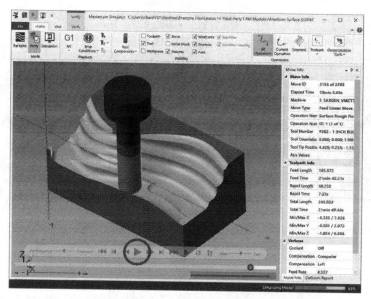

Figure 14.58 The *Mastercam Simulator* window

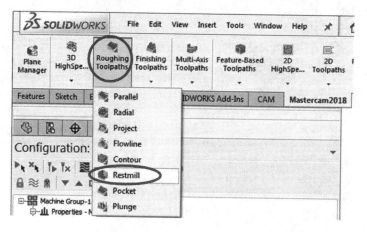

Figure 14.59 Defining *Surface Rough Restmill* operation

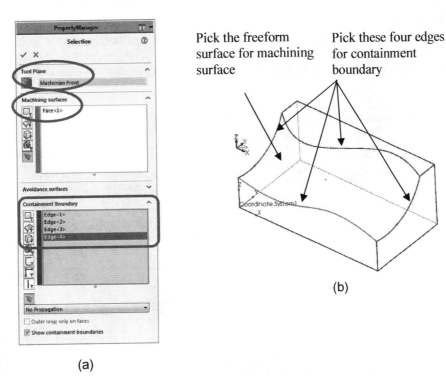

Pick the freeform surface for machining surface

Pick these four edges for containment boundary

(b)

(a)

Figure 14.60 Picking the freeform surface and boundary edges, (a) the *PropertyManager: Selection* dialog box, and (b) picking the freeform surface from the graphics area for machining surface, and picking the four boundary edges of the freeform surface for containment boundary

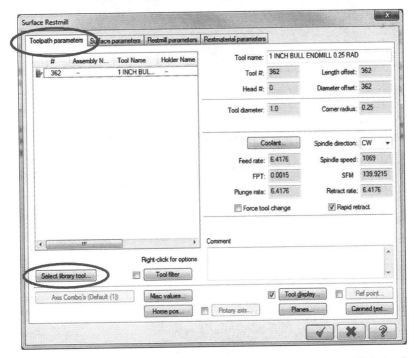

Figure 14.61 Clicking *Select library tool* from the *Surface Restmill* dialog box

We activate the *Containment Boundary* selection by clicking the field (highlighted in blue), circled in Figure 14.60(a). We pick the four boundary edges of the freeform surface for containment boundary consecutively—see Figure 14.60(b), either in CW or CCW order. Click the checkmark ✔ on the top left corner of the *PropertyManager: Selection* dialog box to accept the selection. The *Surface Restmill* dialog box appears (see Figure 14.61).

Under the *Toolpath parameters* tab of the *Surface Restmill* dialog box (Figure 14.61), click *Select library tool* to bring out the *Tool Selection* dialog box (see Figure 14.62), from which we pick Tool #311; 0.5in. ball mill. Click the checkmark ✔ to accept the selection.

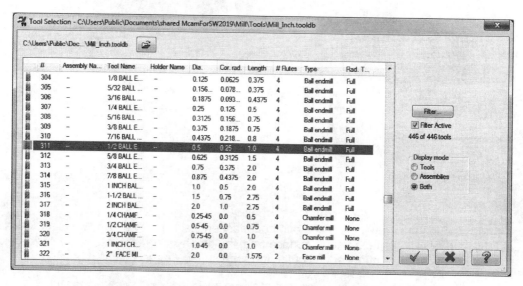

Figure 14.62 Choosing Tool #311 from the *Tool Selection* dialog box

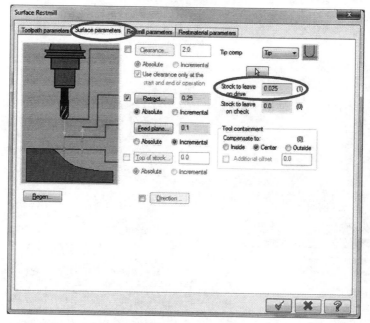

Figure 14.63 Defining stock to leave under the *Surface parameters* tab of the *Surface Restmill* dialog box

Choose the *Surface parameters* tab, and enter 0.025 for *Stock to leave on drive* (see Figure 14.63).

Choose the *Restmill parameters* tab (see Figure 14.64), and enter 0.2 for *Max stepdown*, 0.4 for *Stepover*, and *Zigzag* for *Cutting method* (default).

Choose the *Restmaterial parameters* tab, and choose *All previous operations* for *Compute remaining stock from* (circled in Figure 14.65). Click the checkmark [✓] to accept the definition.

The operation, *2 – Surface Restmill* is listed in the *Toolpaths Manager*, see Figure 14.66(a), and the toolpath is generated; see Figure 14.66(b). Note that toolpaths of both operations, *Surface Rough Flowline* and *Surface Restmill*, are displayed.

You may use the *Toggle display on selected operations* button ≈ , second from the left in the second row above the *Toolpaths Manager*—circled in Figure 14.66(a)—to toggle the display of the toolpath.

Select *1 Surface Rough Flowline*, and click the *Toggle display on selected operations* button ≈ to turn off its toolpath display. Now, the toolpath remaining in the graphics area—see Figure 14.67(a)—is that of the *Surface Restmill* operation.

Select *Toolpath Group-1* in the *Toolpaths Manager*, and click the *Verified selected operations* button 🔧 to carry out a material removal simulation—see Figure 14.67(b)—for combined operations, *Surface Rough Flowline* and *Surface Restmill*, using the *Mastercam Simulator* window.

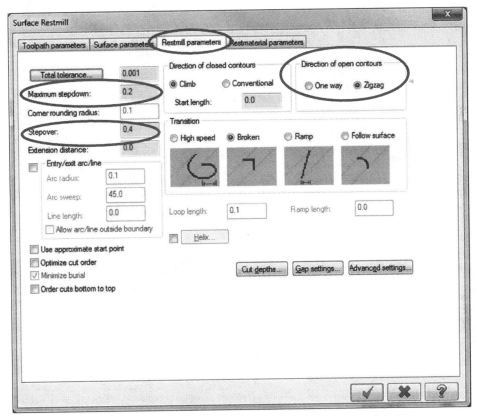

Figure 14.64 Defining cutting parameters under the *Restmill parameters* tab of the *Surface Restmill* dialog box

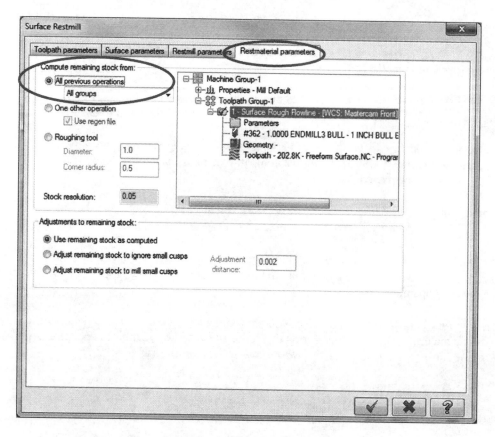

Figure 14.65 Choosing all previous operations under the *Restmaterial parameters* tab of the *Surface Restmill* dialog box

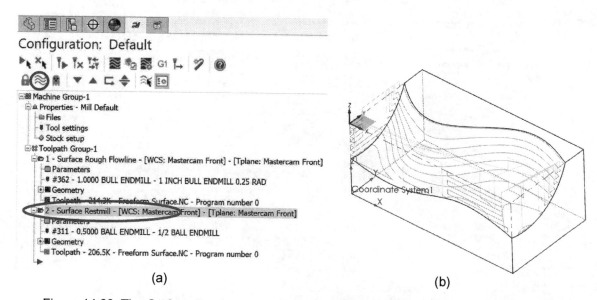

(a)   (b)

Figure 14.66 The *Surface Restmill* operation, (a) *2 – Surface Restmill* listed in the *Toolpaths Manager,* and (b) toolpaths of both operations displayed in the graphics area

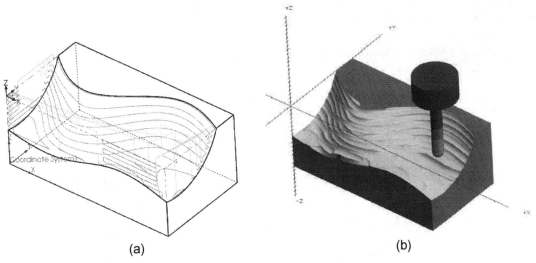

(a)   (b)

Figure 14.67 The *Surface Restmill* operation, (a) toolpath of *Surface Restmill* operation, and (b) material removal simulation for combined operations, *Surface Rough Flowline* and *Surface Restmill*

Multi-Axis Toolpaths: Flowline

Click the *Multi-Axis Toolpaths* command button and choose *Flowline* (circled in Figure 14.68).

In the *Multi-axis Toolpath - Flow* dialog box appearing next (see Figure 14.69) with *Cut Pattern* selected (left column, circled).

Click the arrow button ▨ (circled in Figure 14.69), and pick the freeform surface from the graphics area.

Flowlines along the up-down direction over the freeform surface appear in the graphics area. Similar to what we did earlier, we click *Cut Direction* in the *Flowline options* of the *PropertyManager* dialog box—see Figure 14.56(b)—to toggle the flowlines to be along the longitudinal direction. The flowlines are now along the longitudinal direction, as shown in Figure 14.70(a). Click the checkmark ✓ at the top left corner of the *Flowline* options dialog box to accept the changes.

The number in the parentheses next to the arrow button ▨ is changed from 0 to 1.

Choose *Zigzag* as the *Cutting method*, and enter 0.05 for *Distance* of *Step across*.

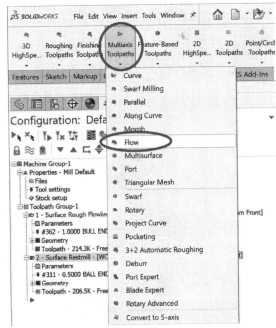

Figure 14.68 Defining *Multi-Axis: Flow* operation

In the *Multi-axis Toolpath - Flow* dialog box, click *Tool Axis Control* (left column, circled in Figure 14.71). Click the drop-down menu next to *Tool axis control* to select *Pattern surface* (circled in Figure 14.71). The freeform surface is automatically selected, and the number in the parentheses next to the arrow button ![arrow button] becomes *1*. Click the drop-down menu next to *Output format* to select *5 axis* (circled in Figure 14.71).

Since we are using the same 0.5in. ball mill, we do not need to select *Tool* on the left column. Click the checkmark ![checkmark] to accept the definition.

Toolpaths will be calculated and displayed in the graphics area. Follow the same steps as before to turn off the toolpath of the *Surface Restmill* operation, and only display the toolpath of the *Multi-Axis Flowline* operation, as shown in Figure 14.70(b).

Click *Toolpath Group-1* in the *Toolpaths Manager*, and click the *Verified selected operations* button ![button] to carry out the material removal simulation for all three operations combined using the *Mastercam Simulator* window (see Figure 14.72).

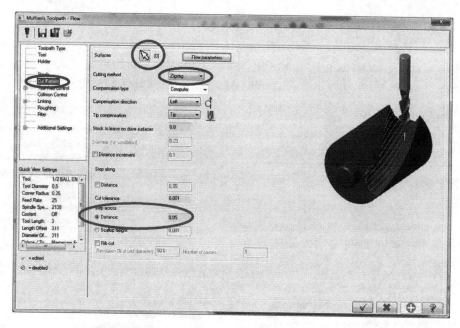

Figure 14.69 Defining *Multi-Axis Flowline* operation

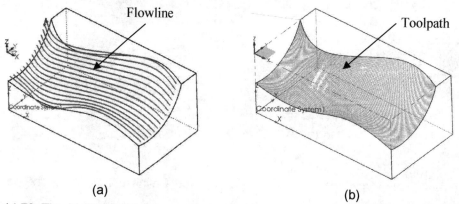

(a)   (b)

Figure 14.70 The *Multi-Axis Flowline* operation, (a) flowlines in the longitudinal direction, and (b) the toolpaths generated

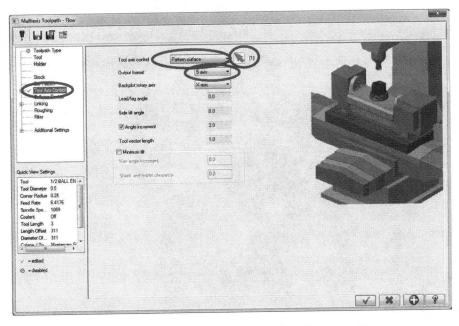

Figure 14.71 Defining *Multi-Axis Flowline* operation

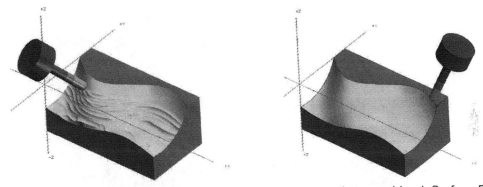

Figure 14.72 The material removal simulation for all three operations combined, Surface Rough Flowline, Surface Restmill Surface, and Multi-Axis Flowline

Mastercam Simulator

Similar to the Machine Simulation of CAMWork*s*, Mastercam Simulator of Mastercam for SOLIDWORKS offers a similar realistic setup in simulating machining operations. For example, a default machine simulator like that of Figure 14.73 provides a setup of tools with a tool holder and stock mounted on a tilt-rotary table.

In the *Mastercam Simulator* window, click the *Simulation* button ⬛ on top, circled in Figure 14.73(a). Click the *Play* button—below the graphics area as circled in Figure 14.73(a)—to simulate the machining operations for the stock sitting on a rotary table of a virtual mill; see a closer view in Figure 14.73(b).

Similar to Machine Simulation of CAMWorks, Mastercam Simulator detects tool collisions when they happen.

We have now completed this exercise. You may close the *Machine Simulation* window, save your model for future reference, and close SOLIDWORKS.

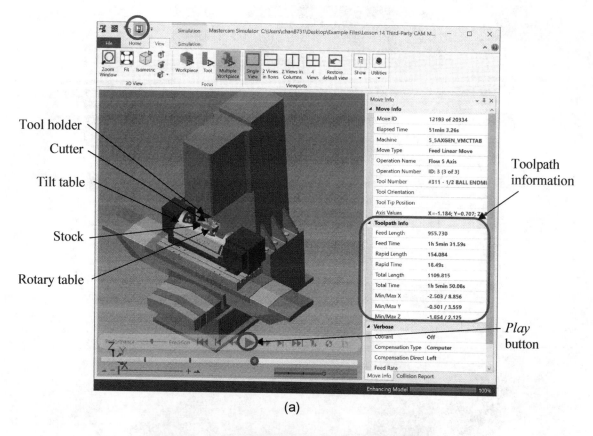

(a)

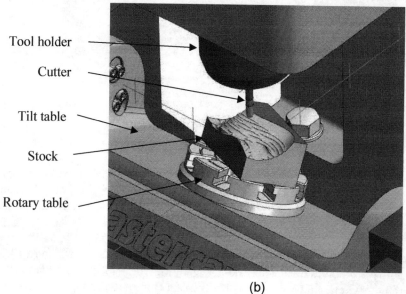

(b)

Figure 14.73 The *Mastercam Simulator* window, (a) the entire user interface window, and (b) the zoom-in view on the work area

14.7 Exercises

Problem 14.1. Follow the same steps discussed in this lesson to machine the same part of Problem 6.1 shown in Figure 14.74 using an aluminum stock of 12in.×5.5in.×3.5in. Report the tools and machining options selected for individual operations. Also report machining times of individual and combined operations. Would operations of 3-axis mill give you a satisfactory machined part? Do you need to create a 5-axis surface milling operation for a satisfactory machined part? You may use CAMWorks, HSMWorks, or Mastercam for SOLIDWORKS of your choice for this exercise.

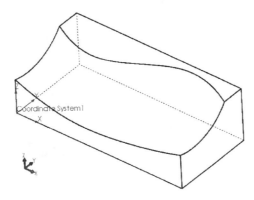

Figure 14.74 Solid model of Problem 14.1

[Notes]

Appendix A: Machinable Features

A.1 Overview

Feature-based machining is the driving concept behind SOLIDWORKS CAM. By defining areas to be machined as features, SOLIDWORKS CAM is able to apply more automation and intelligence into toolpath creation. In this appendix, we discuss machinable features that are extracted by using feature recognition capabilities provided by SOLIDWORKS CAM, including automated feature recognition (AFR), interactive feature recognition (IFR) and local feature recognition (LFR). We focus more on machinable features for milling operations, including the operations generated for the machinable features by the rules, called strategy, implemented in the SOLIDWORKS CAM technology database. We also briefly discuss machinable features for turning operations.

A.2 Machinable Features for Milling Operations

Toolpaths can be generated only on machinable features. SOLIDWORKS CAM provides three methods, AFR, IFR, and LFR, for defining machinable features for milling.

Automatic Feature Recognition (AFR)

Automatic Feature Recognition analyzes the part geometric shape and attempts to extract most common machinable features such as pockets, holes, slots and bosses. You may select the *Extract Machinable Features* button [Extract Machinable Features] above the graphics area, choose from the pull-down menu *Tools > SOLIDWORKS CAM > Extract Machinable Features*, or right click *Mill Part Setup* (under SOLIDWORKS CAM feature tree tab 🔲) and choose *Recognize Features* to initiate AFR. Depending on the complexity of the part, AFR can save considerable time in defining 2.5 axis features, including prismatic features commonly found in milling operations.

Most 2.5 axis features can be extracted automatically. One of the major characteristics of the 2.5 axis features is that the top and bottom faces of the feature are flat and normal to the tool axis of the machining operations, including prismatic solid feature and solid features with tapered wall. Typical 2.5 axis features include boss, pocket, open pocket, corner slot, slot, hole, face feature, open profile, curve or engrave feature. Some of these features are illustrated in Figure A.1.

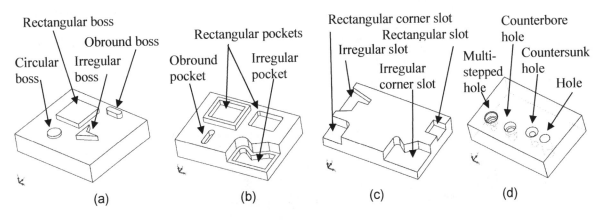

Figure A.1 Illustration of 2.5 axis features, (a) bosses, (b) pockets, (c) slots, and (d) holes

The types of 2.5 axis features to be included in AFR can be selected in SOLIDWORKS CAM. You may choose from the pull-down menu *Tools > SOLIDWORKS CAM > Options* to bring up the *Options* dialog box like that of Figure A.2.

If you click the *Mill Features* tab of the *Options* dialog box, the default feature types to be extracted are listed. By default, holes, non holes (such as pocket, slot, etc.), boss, and tapered & filleted are selected.

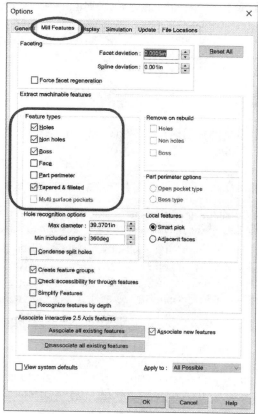

Figure A.2 The *Options* dialog box

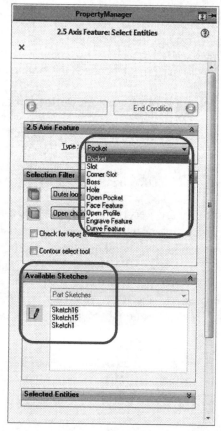

Figure A.3 The *2.5 Axis Features* dialog box

The following feature types are currently supported in SOLIDWORKS CAM:

- Bosses and pockets with vertical walls.
- Bosses and pockets with constant tapered walls with or without constant radius bottom or top fillets and chamfers.
- Pockets and bosses are further broken down into rectangular, circular, obround and irregular.
- Slots with vertical walls.
- Slots have categories for rectangular, corner rectangular, irregular, and corner irregular.
- Numerous hole types including simple holes, counterbores, countersinks, and multi-stepped holes.
- Simple holes can also be described as being drilled, bored, reamed, or threaded.

Local Feature Recognition (LFR)

SOLIDWORKS CAM provides a selective form of Automatic Feature Recognition (AFR) based on user-selected faces in the form of Local Feature Recognition (LFR). This is a semi-automatic method to define features based on face selection. Single or multiple features can be extracted depending on the selected faces.

You may pick one or more faces of the part on which local features are to be recognized. Then, right-click *Mill Part Setup* under the SOLIDWORKS CAM feature tree tab ▥ and select *Recognize Local Features* from the context menu. After executing this command, the locally recognized features, if any, will be added to the machinable feature list.

Interactive Feature Recognition (IFR)

Interactive Feature Recognition allows you to extract interactively either 2.5 axis or multi surface features.

As you may be aware, AFR may not be able to recognize every single feature on complex parts and does not recognize some types of features. To machine these areas, you need to define features manually (or interactively) by right clicking *Mill Part Setup* (under SOLIDWORKS CAM feature tree tab ▥) and choosing, for example, *2.5 Axis Feature*. In the *2.5 Axis Features* dialog box (Figure A.3), select feature type and pick a sketch of the feature to create a machinable feature interactively. Detailed steps for creating such a machinable feature can be found in the book, for example, Lesson 4: A Quick Run Through.

As discussed in Lesson 1, SOLIDWORKS CAM uses a set of knowledge-based rules to assign machining operations to machinable features. The technology database (or TechDB™) contains data and rules that determine operation plans for the respective machinable features. These data and rules can be customized to meet your specific needs. The allowable milling operations for the respective machinable features established in TechDB™ are summarized in Table A.1. Contents of Table A.1 are extracted from SOLIDWORKS CAM Help for your convenience (*Mill > Machinable Features > Automatic Feature Recognition > 2.5 Axis Mill Features Recognized by AFR*).

Table A.1 A list of allowable milling operations for the respective machinable features

Machinable Feature	Description	Allowable Operations
Rectangular Pocket	A pocket whose general shape is a rectangle. The corners of the rectangle can be either sharp or radiused.	Rough Mill, Contour Mill
Circular Pocket	A pocket whose shape is defined by a circle. The difference between a circular pocket and a hole is that a circular pocket can contain an island.	Rough Mill, Contour Mill, Drill, Bore, Ream, Tap, Center Drill, Countersink
Irregular Pocket	A pocket whose shape is neither rectangular nor circular.	Rough Mill, Contour Mill
Machinable Feature	Description	Allowable Operations
Obround Pocket	A rectangular shaped boss with 180 degree round ends.	Rough Mill, Contour Mill

Table A.1 A list of allowable milling operations for the respective machinable features (cont'd)

Machinable Feature	Description	Allowable Operations
Rectangular Slot	A pocket with one edge open to the outside of the part. Machining is extended beyond the open segment of the slot. The general shape of the feature is rectangular. The corners of the rectangle can be either sharp or radiused.	Rough Mill, Contour Mill
Irregular Slot	Similar in definition to a Rectangular Slot except that the general shape is not rectangular.	Rough Mill, Contour Mill
Rectangular Corner Slot	Similar to a Slot except that the feature can contain two or more adjacent edges that are open to the outside of the part. Machining is extended beyond each of these open edges. The corners of the rectangle can be either sharp or radiused.	Rough Mill, Contour Mill
Irregular Corner Slot	Similar to a Rectangular Corner Slot except that the general shape is not rectangular.	Rough Mill, Contour Mill
Rectangular Boss	A Boss whose general shape is rectangular. The corners of the rectangle can be either sharp or radiused.	Contour Mill
Circular Boss	A Boss whose shape is round.	Contour Mill, Thread Mill
Irregular Boss	Any Boss whose shape is neither rectangular nor circular.	Contour Mill
Obround Boss	A rectangular shaped boss with 180 degree round ends.	Contour Mill
Hole	Same shape as a SOLIDWORKS Solids Simple hole with a blind or through end condition or a Simple Drilled hole. Strategies can be assigned as Drill, Bore, Ream, or Thread.	Rough Mill, Contour Mill, Drill, Bore, Ream, Tap, Center Drill, Countersink, Thread Mill
Countersunk Hole	Same shape as a SOLIDWORKS Solids Countersunk hole with a blind or through end condition or a C-Sunk Drilled hole. Strategies can be assigned as Drill, Bore, or Ream.	Rough Mill, Contour Mill, Drill, Bore, Ream, Tap, Center Drill, Countersink, Thread Mill
Counterbored Hole	Same shape as a SOLIDWORKS Solids Counterbored hole with a blind or through end condition or a C-Bored Drilled hole. Strategies can be assigned as Drill, Bore, or Ream.	Rough Mill, Contour Mill, Drill, Bore, Ream, Tap, Center Drill, Countersink, Thread Mill
Countersink/Counterbore Combination	Each countersink/counterbore combination will be recognized as 2 separate hole features that are based on the Mill Part Setup direction.	Rough Mill, Contour Mill, Drill, Bore, Ream, Tap, Center Drill, Countersink, Thread Mill
Multi-stepped Hole	Any hole with multiple steps that is not recognized as a hole, countersunk hole or counterbored hole.	Rough Mill, Contour Mill, Drill, Bore, Ream, Tap, Center Drill, Countersink, Thread Mill

Table A.1 A list of allowable milling operations for the respective machinable features (cont'd)

Machinable Feature	Description	Allowable Operations
Open Pocket	Created only when AFR finds a Boss feature and the machining direction is parallel to one side of the stock. The bottom of the open pocket is the bottom of the boss. The boss becomes an island in the open pocket.	Rough Mill, Contour Mill
Face Feature	If the Face option is checked on the Features tab in the Options dialog box, a Face Feature is created when AFR finds at least one non-face machinable feature, the topmost face is parallel to the Mill Part Setup and the machining direction is parallel to one of the sides of the stock.	Face Mill, Rough Mill, Contour Mill
Perimeter–Open Pocket or Perimeter-Boss Feature	If the Perimeter option is checked on the Features tab in the Options dialog box, AFR creates either a boss or open pocket feature for the perimeter for any Setup created via AFR.	Rough Mill, Contour Mill, Face Mill

A.3 Machinable Features for Turning Operations

Similar to milling operations, machinable features can be extracted in SOLIDWORKS CAM either automatically using Automatic Feature Recognition (AFR) or created interactively by defining a turn feature from a SOLIDWORKS part face, sketch or edge.

The following turn feature types are currently supported:

- Face Feature: A Face feature is defined from vertical edges at the front edge of the part model.
- OD Feature: An OD feature includes the outside shape of the part from the face feature to the cutoff feature, not including the shape of any groove features.
- ID Feature: An ID feature includes the inside diameter shape of the part from the Face feature to the Cut Off feature, not including the shape of any groove features.
- Groove Feature with vertical walls: A Groove is a feature that is closed on both sides and below the surface of the surrounding geometry. There are three categories of grooves: rectangular, half obround and generic. You can cut a groove into the outer diameter, the inner diameter or a face. Groove features where the bottom of the groove is not parallel to the Z or X axis are not recognized by AFR and must be defined interactively.
- Cut Off Feature: A Cut Off feature is defined from vertical edges on the opposite side of the Face feature. A Cut Off feature is similar to a face and can be converted to a Face feature for two-step turning operations using the Convert to Face Feature command on the Cut Off feature context menu.

Some of the aforementioned features are illustrated in Figure A.4. The allowable turning operations for the respective machinable features established in TechDB™ are summarized in Table A.2.

Table A.2 A list of allowable turning operations for the respective machinable features

Machinable Feature	Description	Allowable Operations
Face	A Face feature is defined from vertical edges at the front of the part model. This feature is generally used to trim excess stock off the front of the part.	Face Rough Face Finish

Table A.2 A list of allowable turning operations for the respective machinable features (cont'd)

Machinable Feature	Description	Allowable Operations
OD	An OD (outer diameter) feature includes the outside shape of the part from the face feature to the cutoff feature, excluding the shape of any groove features. Typically, this feature is used to trim the stock away, leaving you with the outer shape of your design.	Turn Rough Turn Finish Face Rough Face Finish Groove Rough Groove Finish Thread
ID	An ID (inner diameter) feature includes the inside diameter shape of the part from the face feature to the cutoff feature, excluding the shape of any groove features.	Bore Rough Bore Finish Drill Thread
Groove Rectangular OD	A Groove Rectangular OD is a recessed feature on the OD of the part where the side walls are equal and parallel to the X machining direction. The bottom and or top may include constant radius fillets or equal size chamfers.	Groove Rough Groove Finish Turn Rough Turn Finish
Groove Rectangular ID	A Groove Rectangular ID is a recessed feature on the ID of the part where the side walls are equal and parallel to the X machining direction. The bottom and or top may include constant radius fillets or equal size chamfers.	Groove Rough Groove Finish Bore Rough Bore Finish
Groove Rectangular Face	A Groove Rectangular Face is a recessed feature on the front edge of the part where the side walls are equal and parallel to the Z machining direction. The bottom and or top may include constant radius fillets or equal size chamfers.	Groove Rough Groove Finish
Groove Half Obround OD	A Groove Half Obround OD is a recessed feature on the OD of the part where the side walls are equal and parallel to the X machining direction. The side walls of the groove are joined at the bottom of the groove by a single 180 degree radius.	Groove Rough Groove Finish
Groove Half Obround ID	A Groove Half Obround ID is a recessed feature on the ID of the part where the side walls are equal and parallel to the X machining direction. The side walls of the groove are joined at the bottom of the groove by a single 180 degree radius.	Groove Rough Groove Finish
Groove Half Obround Face	A Groove Half Obround Face is a recessed feature on the Face of the part where the side walls are equal and parallel to the Z machining direction. The side walls of the groove are joined at the bottom of the groove by a single 180 degree radius.	Groove Rough Groove Finish
Groove Generic OD	A Groove Generic OD is any recessed groove shaped feature on the OD of the part that is neither rectangular nor half obround in shape.	Groove Rough Groove Finish Turn Rough Turn Finish
Groove Generic ID	A Groove Generic ID is any recessed groove shaped feature on the ID of the part that is neither rectangular nor half obround in shape.	Groove Rough Groove Finish Bore Rough Bore Finish
Groove Generic Face	A Groove Generic Face is any recessed groove shaped feature on the face of the part that is neither rectangular nor half obround in shape.	Groove Rough Groove Finish
Cut Off	A Cut Off feature is defined from vertical edges on the opposite side of a Face feature. This feature is used primarily to trim excess stock from the back of the part. A Cut Off feature is similar to a face and can be converted to a Face feature for two-step turning operations using the Convert to Face Feature command on the Cut Off feature context menu.	Cut Off Face Rough Face Finish Groove Rough Groove Finish

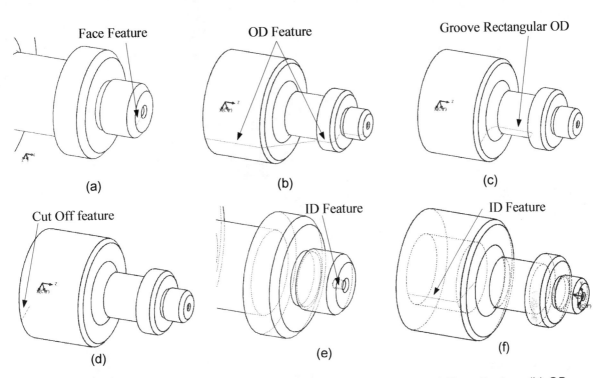

Figure A.4 Examples of machinable features for turning operations, (a) *Face Feature*, (b) *OD Feature*, (c) *Groove Rectangular OD*, (d) *Cut Off Feature*, (e) *ID Feature* (front end hole), and (f) *ID Feature* (rear end hole)

[Notes]

Appendix B: Machining Operations

B.1   Overview

When a machinable feature is extracted or manually created, the corresponding machining operations are generated by the knowledge rules and data stored in the TechDB™. In this appendix, we discuss briefly the operations involved in both milling and turning operations.

B.2   Milling Operations

Milling operations are grouped by the types of machinable features, i.e., 2.5 axis features and 3 axis cutting cycles.

2.5 Axis Features

The operations that support machining 2.5 axis features include Rough Mill, Contour Mill, Face Mill, Thread Mill, and Single Point.

A Rough Mill operation removes material from a part by following the shape of the machinable feature (pocket-in, pocket-out, spiral in or spiral out) or by making a series of parallel cuts across the machinable feature (zigzag, zig or plunge rough).

A Contour Mill operation removes material from a part by following the shape of the profile of pockets, slots, bosses, etc.

The Face Mill operation generates a toolpath on a mill Face feature for squaring or facing off the top of a part. Although a Rough Mill operation can generate toolpath that will remove material from a Face feature, the Face Mill operation provides specific controls to produce a toolpath motion that is more appropriate for this task.

Thread Mill operations create thread mill toolpaths for a hole or circular boss. A thread mill operation can be generated automatically by assigning a thread mill strategy to a hole or circular boss and selecting Generate Operation Plan. Alternatively, you can insert a thread mill operation using the New Operation and New Hole Operation commands.

Single point cycles include Drill, Bore, Ream, Tap, Countersink, and Centerdrill.

3-Axis Cutting Cycles

The 3 Axis cutting cycles include Area Clearance, Z Level, and Flat Area.

The Area Clearance cycle removes the material between the stock or contain area and the selected feature at decreasing Z depth levels by making a series of parallel cuts across the stock (Lace) or by pocketing out toward the stock.

The Z Level cycle is a finish contouring cycle that removes material by making a series of horizontal planar cuts. The cuts follow the contour of the feature at decreasing Z levels based on the Surface Finish you specify. Cutting starts from the highest location on the model and works down.

The Flat Area cycle uses a pocket out pattern to remove material from feature faces that are flat and parallel to the XY machining plane. SOLIDWORKS CAM generates toolpaths only on completely flat areas. If a face/surface has even a small gradient, SOLIDWORKS CAM will not generate a toolpath. This cycle can be used for finishing where excess material has already been cleared and supports single or multiple depths of cut.

B.3   Turning Operations

The Turn cutting cycles include Face Rough, Face Finish, Turn Rough, Turn Finish, Groove Rough, Groove Finish, Bore Rough, Bore Finish, Drill, Center Drill, Thread, and Cut Off.

Face Rough: The Face Rough Operation defines multiple rough cuts to machine turned faces.

Face Finish: The Face Finish Operation defines a single finish cut to machine turned faces.

Turn Rough: The Turn Rough operation defines multiple rough cuts to machine turned faces. This is a common machining operation for a part that has a range of stock on the ODs that requires several cuts to remove the bulk of the stock.

Turn Finish: The Turn Finish operation defines multiple rough cuts to machine turned faces.

Groove Rough/Finish: The Grooving Cycle (Rough and Finish) allows you to cut a groove of almost any shape.

Bore Rough/Finish: The Bore Cycle (Rough and Finish) allows you to bore a hole or an ID feature.

Drill: The Drill operation generates a drill toolpath at the center line of the part (Z axis).

Center Drill: The Center Drill operation generates a center drill toolpath at the center point of the hole with a small depth.

Thread: In SOLIDWORKS CAM Turning, you use Strategies to define a thread on a feature. In order to generate a Thread operation, the corresponding thread condition must be selected from the TechDB™. The major diameter for the thread condition in the TechDB™ must match the feature's maximum diameter.

Cut off: When the stock is defined as bar stock, Automatic Feature Recognition generates a Cut off feature on the opposite side of the Face feature. A Cut off operation is generated for a Cut off feature when you generate an Operation Plan.

Appendix C: Alphabetical Address Codes

The alphabetical address codes listed in Table C.1 are extracted from HAAS operation manual.

Table C.1 Alphabetical Address Codes for HAAS CNC Machines

A	Fourth axis rotary motion	The A address character is used to specify motion for the optional fourth, A, axis. It specifies an angle in degrees for the rotary axis. It is always followed by a signed number and up to three fractional decimal positions. If no decimal point is entered, the last digit is assumed to be 1/1000 degrees. The smallest magnitude is 0.001 degrees, the most negative value is -99999.000 degrees, and the largest number is 99999.000 degrees.
B	Fifth axis rotary motion	The B address character is used to specify motion for the optional fifth, B, axis. It specifies an angle in degrees or the rotary axis. It is always followed by a signed number and up to three fractional decimal positions. If no decimal point is entered, the last digit is assumed to be 1/1000 degrees. The smallest magnitude is 0.001 degrees, the most negative value is -8380.000 degrees, and the largest number is 8380.000 degrees
C	Auxiliary external rotary axis	The C address character is used to specify motion for the optional external sixth, C, axis. It specifies an angle in degrees for the rotary axis. It is always followed by a signed number and up to three fractional decimal positions. If no decimal point is entered, the last digit is assumed to be 1/1000 degrees. The smallest magnitude is 0.001 degrees, the most negative value is -8380.000 degrees, and the largest number is 8380.000 degrees.
D	Tool diameter selection	The D address character is used to select the tool diameter or radius used for cutter compensation. The number following must be between 0 and 100. D0 specifies that the tool size is zero and serves to cancel a previous Dn. Any other value of D selects the numbered entry from the tool diameter/radius list under the Offsets display.
E	Contouring accuracy	The E address character is used, with G187, to select the accuracy required when cutting a comer during high speed machining operations. The range of values possible for the E code is 0.0001 to 0.25. Refer to the "Contouring Accuracy" section for more information.
F	Feed rate	The F address character is used to select the feed rate applied to any interpolation functions, including pocket milling and canned cycles. It is either in inches per minute with four fractional positions or mm per minute with three fractional positions. When G93 (Inverse Time) is programmed, F is in blocks per minute, up to a maximum of 15400.0000 inches per minute (39300.000 millimeters per minute).
G	Preparatory functions (G codes)	The G address character is used to specify the type of operation to occur in a block. The G is followed by a two or three digit number between 00 and 150. Each G code is part of a numbered group. The Group 0 codes are non-modal; that is, they specify a function applicable to this block only and do not affect other blocks. The other groups are modal and the specification of one code in the group cancels the previous code applicable from that group. A modal G code applies to all subsequent blocks so those blocks do not need to re-specify the same G code. Multiple G codes can be placed in a block in order to specify all of the setup conditions for an operation, provided no two are from the same numbered group.
H	Tool length offset selection	The H address character is used to select the tool length offset entry from the offsets memory. The H is followed by a two digit number between 0 and 100. H0 will cause no offset to be used and Hnn will use the tool length entry n from the Offsets display. Note that G49 is the default condition and will clear the tool length offsets; so you must select either G43 or G44 to activate tool length offsets. The TOOL OFSET MESUR button will enter a value into the offsets to correspond to the use of G43.

Table C.1 Alphabetical Address Codes for HAAS CNC Machines (cont'd)

I, J, K	Canned cycle and circular optional data	The I address character is used to specify data used for some canned cycles and circular motions. It is either n inches with four fractional positions or mm with three fractional positions. It is followed by a signed number in inches between -15400.0000 and 15400.0000 or between -39300.000 and 39300.000mm for metric.
		The J address character is used to specify data used for some canned cycles and circular motions. It is formatted just like the I data. It is followed by a signed number in inches between -15400.0000 and 15400.0000 or between -39300.000 and 39300.000mm for metric.
		The K address character is used to specify data used for some canned cycles and circular motions. It is formatted just like the I data. It is followed by a signed number in inches between -15400.0000 and 15400.0000 or between -39300.000 and 39300.000mm for metric.
L	Loop count for repeated cycles	The L address character is used to specify a repetition count for some canned cycles and auxiliary functions. It is followed by an unsigned number between 0 and 32767.
M	Code miscellaneous functions	The M address character is used to specify an M code for a block. These codes are used to control miscellaneous machine functions. Note that only one M code is allowed per block of the CNC program and all M codes are performed at the end of the block.
N	Number of block	The N address character is entirely optional. It can be used to identify or number each block of a program. It is followed by a number between 0 and 99999. The M97 function must reference an N line number.
O	Program number/name	The O address character is used to identify a program. It is followed by a number between 0 and 9999. A program saved in memory always has an Onnnnn identification in the first block; it cannot be deleted. Altering the O in the first block causes the program to be renamed. An Onnnnn can be placed in other blocks of a program but will have no effect and can be confusing to the reader. A colon (:) may be used in the place of O in a program but is always displayed as "O".
P	Delay time, scaling factor, or program number	The P address character is used to enter either a time in seconds or a program number for a subroutine call. If it is used as a time (for a G04 dwell), it may be a positive decimal with fraction between 0.001 and 1000.0. If it is used as a program name (for an M98), or a line number (for an M97), the value may be a positive number without decimal point up to 9999. It can also be a scaling factor (G51).
Q	Canned cycle optional data	The Q address character is used in canned cycles and is followed by a signed number in inches between 0 and 8380.000 for inches or between 0 and 83800.00 for metric.
R	Canned cycle and circular optional data	The R address character is used in canned cycles and circular interpolation. It is either in inches with four fractional positions or mm with three fractional positions. It is followed by a signed number in inches between -15400.0000 and 15400.0000 for inches or between -39300.000 and 39300.000 for millimeters. It is usually used to define the reference plane for canned cycles.
S	Spindle speed command	The S address character is used to specify the spindle speed in conjunction with M41 and M42. The S is followed by an unsigned number between 1-99999. The S command does not turn the spindle on or off; it only sets the desired speed. If a gear change is required in order to set the commanded speed, this command will cause a gear change to occur even if the spindle is stopped. If the spindle is running, a gear change operation will occur and the spindle will continue running at the new speed.
T	Tool selection code	The T address character is used to select the tool for the next tool change. The number following must be a positive number between 1 and the number in Parameter 65. It does not cause the tool change operation to occur. The Tn may be placed in the same block that starts the tool change (M6 or M16) or in any previous block.

Table C.1 Alphabetical Address Codes for HAAS CNC Machines (cont'd)

U, V, W	Auxiliary external linear axis	The U, V, and W address character are used to specify motion for the optional external linear, U, V, and W axes. It specifies a position of motion. It is either in inches with three or four fractional positions or millimeters with three fractional positions. It is followed by a signed number between -838.0000 and 838.0000 or -8380.000 and 8380.000 or in millimeters between -8380.000 and 8380.000. If no decimal point is entered, the last digit is assumed to be 1/1000 inches or 1/1000mm.
X, Y, Z	Linear X-, Y-, and Z-axis motions	The X, Y, and Z address character are used to specify motion for the X-axis. It specifies a position or distance along the X-, Y-, and Z-axes, respectively. It is either in inches with four fractional positions or mm with three fractional positions. It is followed by a signed number in inches between -15400.0000 and 15400.0000 for inches or between -39300.000 and 39300.000 for millimeters. If no decimal point is entered, the last digit is assumed to be 1/10000 inches or 1/1000 mm.

[Notes]

Appendix D: Preparatory Functions

The Preparatory Functions or G codes listed in Table D.1 are extracted from HAAS operation manual.

Table D.1 G codes of HAAS CNC machines

Code:	Group	Function:	Code:	Group	Function:
G00	01	*Rapid Motion	G76	09	Fine Boring Canned Cycle
G01	01	Linear Interpolation Motion	G77	09	Back Bore Canned Cycle
G02	01	CW Interpolation Motion	G80	09	* Canned Cycle Cancel
G03	01	CCW Interpolation Motion	G81	09	Drill Canned Cycle
G04	00	Dwell	G82	09	Spot Drill Canned Cycle
G09	00	Exact Stop	G83	09	Normal Peck Drill Canned Cycle
G10	00	Set Offsets	G84	09	Tapping Canned Cycle
G12	00	CW Circular Pocket Milling (Yasnac)	G85	09	Boring Canned Cycle
G13	00	CCW Circular Pocket Milling (Yasnac)	G86	09	Bore/Stop Canned Cycle
G17	02	* XY Plane Selection	G87	09	Bore/Stop/Manual Retract Canned Cycle
G18	02	ZX Plane Selection	G88	09	Bore/Dwell/Manual Retract Canned Cycle
G19	02	YZ Plane Selection	G89	09	Bore/ Dwell Canned Cycle
G20	06	Select Inches	G90	03	* Absolute
G21	06	Select Metric	G91	03	Incremental
G28	00	Return To Reference Point	G92	00	Set Work Coordinates -FANUC or HMS
G29	00	Return From Reference Point	G92	00	Set Work Coordinates -YASNAC
G31	00	Feed Until Skip (optional)	G93	05	Inverse Time Feed Mode
G35	00	Automatic Tool Diameter Measurement (optional)	G94	05	Feed Per Minute Mode
G36	00	Automatic Work Offset Measurement (optional)	G98	10	* Initial Point Return
G37	00	Automatic Tool Offset Measurement (optional)	G99	10	R Plane Return
G40	07	* Cutter Comp Cancel	G100	00	Cancel Mirror Image
G41	07	2D Cutter Compensation Left	G101	00	Enable Mirror Image
G42	07	2D Cutter Compensation Right	G102	00	Programmable Output To RS-232
G43	08	Tool Length Compensation +	G103	00	Limit Block Buffering
G44	08	Tool Length Compensation -	G107	00	Cylindrical Mapping
G47	00	Text Engraving	G110	12	Select Work Coordinate System 7
G49	08	* G43/G44/G143 Cancel	G111	12	Select Work Coordinate System 8
G50	11	G51 Cancel	G112	12	Select Work Coordinate System 9
G51	11	Scaling (optional)	G113	12	Select Work Coordinate System 10
G52	12	Set Work Coordinate System G52 (Yasnac)	G114	12	Select Work Coordinate System 11
G52	00	Set Local Coordinate System (Fanuc)	G115	12	Select Work Coordinate System 12
G52	00	Set Local Coordinate System (HMS)	G116	12	Select Work Coordinate System 13
G53	00	Non-Modal Machine Coordinate Selection	G117	12	Select Work Coordinate System 14
G54	12	* Select Work Coordinate System 1	G118	12	Select Work Coordinate System 15
G55	12	Select Work Coordinate System 2	G119	12	Select Work Coordinate System 16
G56	12	Select Work Coordinate System 3	G120	12	Select Work Coordinate System 17
G57	12	Select Work Coordinate System 4	G121	12	Select Work Coordinate System 18
G58	12	Select Work Coordinate System 5	G122	12	Select Work Coordinate System 19
G59	12	Select Work Coordinate System 6	G123	12	Select Work Coordinate System 20
G60	00	Unidirectional Positioning	G124	12	Select Work Coordinate System 21
G61	13	Exact Stop Modal	G125	12	Select Work Coordinate System 22
G64	13	* G61 Cancel	G126	12	Select Work Coordinate System 23
G65	00	Macro Subroutine Call (optional)	G127	12	Select Work Coordinate System 24
G68	16	Rotation (optional)	G128	12	Select Work Coordinate System 25
G69	16	G68 Cancel (optional)	G129	12	Select Work Coordinate System 26
G70	00	Bolt Hole Circle (Yasnac)	G136	00	Automatic Work Offset Center Measurement
G71	00	Bolt HoleArc (Yasnac)	G141	07	3D+ Cutter Compensation
G72	00	Bolt Holes Along an Angle (Yasnac)	G143	08	5 AX Tool Length Compensation (optional)
G73	09	High Speed Peck Drill Canned Cycle	G150	00	General Purpose Pocket Milling
G74	09	Reverse Tap Canned Cycle	G174/184	00	General-Purpose Rigid Tapping (optional)

* Default

[Notes]

Appendix E: Machine Functions

The Machine Functions or M codes listed in Table E.1 are extracted from HAAS operation manual.

Table E.1 M codes of HAAS CNC machines

Code:	Function:
M00	Stop Program
M01	Optional Program Stop
M02	Program End
M03	Spindle Forward
M04	Spindle Reverse
M05	Spindle Stop
M06	Tool Change
M08	Coolant On
M09	Coolant Off
M10	Engage 4th Axis Brake
M11	Release 4th Axis Brake
M12	Engage 5th Axis Brake
M13	Release 5th Axis Brake
M16	Tool Change (same as M06)
M19	Orient Spindle
M21-M24	Optional Pulsed User M Function with Fin
M30	Prog End and Rewind
M31	Chip Conveyor Forward
M32	Chip Conveyor Reverse
M33	Chip Conveyor Stop
M34	Increment Coolant Spigot Position
M35	Decrement Coolant Spigot Position
M36	Pallet Rotate
M39	Rotate Tool Turret
M41	Low Gear Override
M42	High Gear Override
M51-M54	Set Optional User M
M61-M64	Clear Optional User M
M75	Set Measure point
M76	Disable Displays
M77	Enable Displays
M78	Alarm if skip signal found
M79	Alarm if skip signal not found
M82	Tool Unclamp
M86	Tool Clamp
M88	Through the Spindle Coolant ON
M89	Through the Spindle Coolant OFF
M95	Sleep Mode
M96	Jump if no Input
M97	Local Sub-Program Call
M98	Sub Program Call
M99	Sub Program Return Or Loop

[Notes]